巴蜀文化国际传播研究中心资助项目（项目编号：2023YB13）

大熊猫列国游记

Journey to the World of Giant Pandas

李守林　付洪涛◎著

西南交通大学出版社
·成　都·

图书在版编目（CIP）数据

大熊猫列国游记 = Journey to the World of Giant Pandas : 汉文、英文 / 李守林，付洪涛著. -- 成都 : 西南交通大学出版社，2025. 3. -- ISBN 978-7-5643-9993-1

Ⅰ. Q959.838-49

中国国家版本馆 CIP 数据核字第 2024QN5891 号

Daxiongmao Lieguo Youji

大熊猫列国游记

Journey to the World of Giant Pandas

李守林　付洪涛　著

策划编辑	胡　军
责任编辑	梁　红
封面设计	曹天擎
出版发行	西南交通大学出版社 （四川省成都市金牛区二环路北一段 111 号 西南交通大学创新大厦 21 楼）
营销部电话	028-87600564　028-87600533
邮政编码	610031
网　址	http://www.xnjdcbs.com
印　刷	四川玖艺呈现印刷有限公司
成品尺寸	185 mm × 260 mm
印　张	11.75
字　数	185 千
版　次	2025 年 3 月第 1 版
印　次	2025 年 3 月第 1 次
书　号	ISBN 978-7-5643-9993-1
定　价	58.00 元

序言

研究表明，大熊猫已在地球上生存了至少800万年，被誉为冰川时代“动物活化石”，是世界自然基金会的形象大使、世界生物多样性保护的旗舰物种、中国特有的物种，主要栖息在中国四川、陕西和甘肃的山林里。

长期以来，中国人与大熊猫友好相处，视大熊猫为神兽、瑞兽，并在20世纪80年代以前将它作为国礼赠送给友好国家。据《日本皇家年鉴》记载，早在唐朝时，中国就曾送给日本天武天皇两只大熊猫，开启了大熊猫国际交流的先河。

1869年以后，一批又一批的西方探险家、游猎家和博物馆标本采集者来到大熊猫产区，试图得到这种珍奇的动物并揭开大熊猫之谜。大熊猫被他们带到欧美后，整个西方，上自王室，下至平民，无不对这一来自东方的“神秘物种”充满好奇，都想看看它们的样子。

中华人民共和国成立后，大熊猫开始扮演“友好大使”角色，在增进中国与外国的友谊和相互了解方面发挥了重要作用。无论走到哪里，大熊猫都会触动人们的心弦。他们已成为世界和平与友谊的使者，并将继续为促进东西方文化交流发挥重大作用。

这是一本介绍大熊猫进入人类世界的书，本书旨在倡导人与自然和谐共处。但历史上，总有一些人，因为私欲，违背自然规律，破坏大自然的生态平衡，例如书中提到的美国人露丝对中国独有的大熊猫产生了强烈的欲望，数次从中国偷走大熊猫，以及罗斯福兄弟组织一支探险队前往中国猎杀大熊猫。他们的行为，对大熊猫造成了极大的伤害，影响恶劣。中华人民共和国成立前，汶川、宝兴、平武等地有数百只大熊猫被猎捕后送往欧美展出。大量的捕捉导致这些地区的大熊猫品种群结构被破坏，加之大熊猫种群密度低、繁殖力低下，迄今当地的大熊猫种群仍未完全恢复。中华人民共和国成立后，中国政府非常重视大熊猫的保护工作，在保护大熊猫方面做出了巨大的努力，取得了巨大的成就。可以说，大熊猫在促进国际交流，促进人与自然和谐共生方面做出了巨大贡献。

Preface

Studies show that giant pandas have been living on the earth for at least 8 million years. Known as the "Living Animal Fossil" of the ice age, the image ambassador of World Wildlife Fund and the flagship species of world biodiversity conservation, the giant panda is endemic to China, mainly living in the mountains of Sichuan, Shaanxi and Gansu provinces in China.

For a long time, the Chinese people have got along well with giant pandas, treating them as a mythical and auspicious beast, and giving it to friendly countries as a national gift before the 1980s. According to the *Japanese Royal Yearbook*, as early as the Tang Dynasty, two giant pandas were given to the Japanese Emperor Tianwu and therefrom started the international communication of giant pandas.

Since 1869, groups of western explorers, safari hunters and museum specimen collectors had come to the giant panda region to try to obtain this rare animal and uncover the mystery of the giant panda. Many giant pandas were brought back to the west. From the royal family to the common people, all were eager to see the "mysterious species" from the East.

After the founding of the People's Republic of China, the giant panda began to play the role of "ambassador of friendship" and played an important role in enhancing the friendship and mutual understanding between China and foreign countries. Wherever they go, these pandas will touch people's hearts. They have become ambassadors of world peace and friendship and will continue to play a significant role in bridging the cultural gap between the East and the West.

This book introduced how the giant panda entered the human world and is designed to advocate for the harmonious coexistence between man and nature. But in history, there were some people who, because of selfish desires, violated the laws of nature and destroyed the ecological balance of nature. For example, mentioned in the book, Ruth had a strong desire for the unique Chinese pandas and stole the pandas from China several times, and the Roosevelt Brothers organized an expedition to China to shot pandas. They caused great harm to the pandas. Before the founding of the People's Republic of China, hundreds of giant pandas in Wenchuan, Baoxing, Pingwu and other places were hunted and sent to Europe and the United States for exhibition. The large-scale hunting had destructed the structure of the giant panda species in these areas, coupled with the low population density and low fecundity, so far the local giant panda population has not fully recovered. Since the founding of the People's Republic of China, the Chinese government has been attaching great importance to the protection of giant pandas, and has made great achievements in protecting them. It can be said that the giant panda has made a great contribution to promoting international communication and promoting the harmonious coexistence between man and nature.

第二篇　至尊国礼

Chapter II State Gift / 025

第一篇
奇珍异兽

几百万年来，大熊猫隐身于中国西南的山林之间，与当地人民和谐共生，优哉游哉。然而，1869年以后，它们的隐秘生活发生了翻天覆地的变化。一批又一批西方探险家、游猎家和博物馆标本采集者远涉万里来到中国，只为得到它们。

Chapter I Rare Species

For millions of years, pandas had been hidden in the mountains and forests of southwest China, harmonious with the local people, leisurely and carefree. However, their hidden lives have changed dramatically since 1869. A group of Western explorers, safaris hunters and museum specimen collectors went to lengths in China just to get them.

异兽出山林
Rare Species Came out of Forest

1862年，法国博物学家阿尔芒·戴维受天主教遣使会派遣，到中国为法国国家自然历史博物馆收集资料，主要是动物学资料。在得知偏居四川西部的穆坪（今宝兴县）境内动物物种丰富后，戴维于1869年来到了宝兴。此时的他还不知道他将与大熊猫结下终生不解之缘并因此闻名世界。

阿尔芒·戴维（一）
Abbe Armand David

In 1862, the French naturalist Abbe Armand David was sent by the Catholic Congrégation de Mission to China, to collect materials for French National Natural History Museum (MNHN), mainly zoological materials. Learning that Muping (now Baoxing County), located in western Sichuan Province, is rich in animal species, David came to Baoxing in 1869, when he didn't know that he would form an indissoluble bond with the giant panda and became famous for it.

1869年3月11日，在外出考察返回途中，戴维发现一户农家挂着一张他从未见过的黑白兽皮。4月1日，戴维得到一只活体成年“白熊”，并给它取名为“黑白熊”。他决定将它带回法国，遗憾的是“黑白熊”还没抵达成都就死了。戴维把它做成标本送回法国国家自然历史博物馆展出。这是“竹林隐士”大熊猫在俗世的第一次公开亮相，也是现代科学意义上的首次发现。

On March 11, 1869, in a farmer's house on the way back from an investigation, David found a black and white fur, which he had never seen before. On April 1, David got a live adult "white bear" and named it Ursus Melanoleucus A. D. (Black and White Bear). He decided to take it back to France. Unexpectedly, the "black and white bear" died before arriving in Chengdu. David made it into a specimen and sent it back to MNHN for display. It is the first public appearance of the giant panda from its "Hermit" life in bamboo forest and the first discovery in the modern scientific sense.

阿尔芒·戴维（二）
Abbe Armand David

世界上第一具大熊猫标本
The first specimen of the giant panda in the world

随着阿尔芒·戴维在中国的惊世发现以及大熊猫标本在巴黎的展出，中国有大熊猫的消息很快传遍西方，引起了国际生物学界的轰动。一批又一批西方探险家、游猎家和博物馆标本采集者来到中国，试图得到这种珍奇的动物。

With the stunning discovery of Armand David in China and the exhibition of the giant panda specimens in Paris, the news that the giant panda appeared in China quickly spread around the west world, causing a sensation in the international biology community. Groups of western explorers, safari and museum species collectors came to China, wishing to capture the exotic animals.

1928年年底，美国第26任总统西奥多·罗斯福的两个儿子——小西奥多·罗斯福和克米特·罗斯福在芝加哥“菲尔德自然历史博物馆”的资助下，组织探险队到中国寻找大熊猫。他们到达中国后就直奔四川省宝兴县——大熊猫发现地。他们在此搜寻了半个多月却一无所获。

In late 1928, the Roosevelt brothers, supported by the Field Museum of Natural History in Chicago, organized an expedition to hunt giant pandas in China. They entered China, and went straight to Baoxing, Sichuan. They searched there for more than half a month, but found nothing.

不得已，他们又转道芦山、荥经、汉源去碰运气。1929年4月13日，罗斯福兄弟俩在原雅安市石棉县擦罗乡发现了一只大熊猫，经过两个多小时的追踪，最后在凉山州和雅安市交界的冕宁县冶勒乡猎杀了这只大熊猫，并将它做成标本带回美国费城博物馆。

They had to turn to Lushan, Yingjing, Hanyuan to try their luck. On April 13, 1929, the Roosevelt brothers found a giant panda in Caluo Town, Shimian County, Ya'an City. After more than two hours of trailing, they finally shot the giant panda in Yele Town, Mianning County, at the border of Liangshan Prefecture and Ya'an City, and made it a specimen and took it back to the Philadelphia Museum.

此时，志得意满的罗斯福兄弟还没有意识到人与自然和谐相处的重要性。多年以后，当怀抱着活泼可爱的大熊猫幼崽"苏琳"的时候，小西奥多·罗斯福才突然意识到自己曾经给大熊猫造成的伤害，他忏悔道："如果要把这个小家伙当作我枪下的纪念品，那我宁可用我的小儿子来代替。人类的好奇心竟会引导犯罪，我再也不会射杀大熊猫了。"[①]

At that time, the proud Roosevelt brothers did not realize the importance of harmony between man and nature. Years later, when holding the lively and lovely giant panda cub Su-Lin, Theodore Roosevelt suddenly realized the damage he had done to the giant panda, and he repented: "If this little guy were to be a souvenir under my gun, I would rather it's my little son instead. Human curiosity can lead to crime, and I shall never shoot any panda again."

①高富华：《大熊猫史话1869—2019》，四川民族出版社2019年版，第102页。

“婴熊”入江湖

Giant Panda Cub Debuted the World

1936年4月17日，美国服装设计师露丝·哈克尼斯前往中国寻找大熊猫。11月9日，露丝在汶川县得到一只大熊猫幼崽并给它取名“苏琳”。最终，露丝将“苏琳”带回了美国。“苏琳”成为现代科学史上第一只走进公众视野的活体大熊猫。

On April 17, 1936, Ruth Harkness went to China to hunt giant pandas. On November 9, Ruth obtained a giant panda cub in Wenchuan County and named it Su-Lin. Ruth brought Su-Lin to the United States alive at last. Su-Lin therefore became the the first giant panda to enter the public view alive in the history of modern science.

露丝与“苏琳”
Ruth with Su-Lin

露丝向美国公众展示“苏琳”
Ruth shows off Su-Lin to the Americans

1936年12月18日，“苏琳”抵达旧金山，随后被送到各大城市展出，所到之处无不引起轰动。诸多社会名流包括秀兰·邓波儿、海伦·凯勒、罗斯福兄弟以及迪安·赛奇等均前往参观。1937年2月8日，“苏琳”定居芝加哥动物园，深受美国人民喜爱。1938年4月1日，“苏琳”死于肺炎，后被制成标本存放在芝加哥菲尔德自然历史博物馆。

On December 18, 1936, the infant Su-Lin arrived in San Francisco and then was brought to many cities to begin the international obsession with pandas. Many celebrities came to visit it, including Shirley Temple, Helen Keller, the Roosevelt brothers and Dean Sage. On February 8, 1937, Su-Lin's odyssey ended at the Chicago Zoological Park. Loved by the American people, Su-Lin died of pneumonia on April 1, 1938, and was made into specimen and stored at the Field Museum of Natural History, Chicago.

在芝加哥动物园的资助下，1937年露丝第二次到中国猎捕大熊猫。11月19日，露丝得到了一只大熊猫幼崽“阴”，但没过多久“阴”就莫名死了。12月18日，露丝又得到了一只熊猫幼崽，露丝为其取名“戴安娜”。1938年2月18日，露丝将其带到芝加哥并将其改名为“妹妹”。1942年8月3日，“妹妹”的生命也走到了尽头。

With funding from the Chicago Zoological Park, Ruth went to China again in 1937. On November 19, Ruth got a giant panda cub Yin, which soon died inexplicably. On December 18, Ruth got another panda cub, which was named Diana. Ruth took the cub to Chicago on February 18, 1938 and later renamed it Mei-Mei, which died on August 3, 1942.

露丝带“妹妹”与“苏琳”（右）见面
Ruth takes Mei-Mei to meet Su-Lin(R)

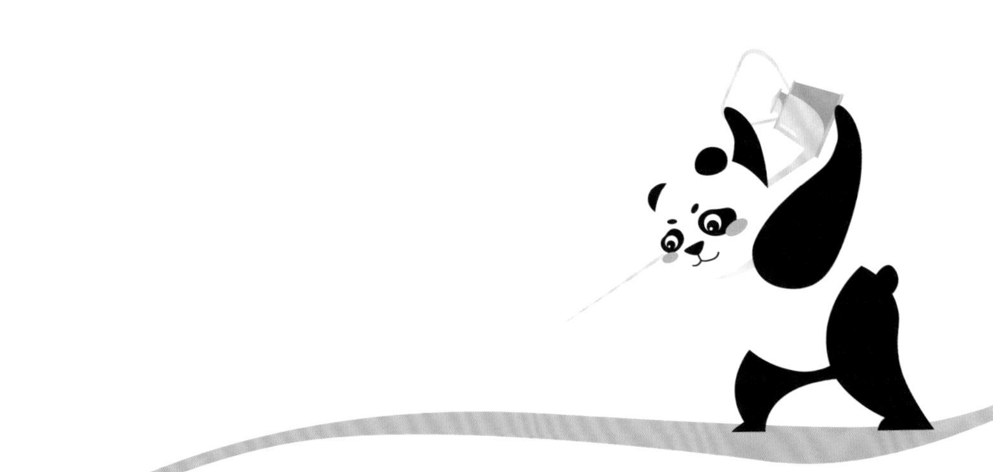

1938年6月初，露丝第三次来到中国并再一次得到一只大熊猫。她给大熊猫取名为“美龄”，几天后又给其改名为“欧琳”。最后，她又将其名字改为“苏森”。

In early June 1938, Ruth came to China for the third time and got a giant panda. She named the panda Mei-Ling, and a few days later, she named it O-Lin. Finally, she changed it’s name to Su-Sen.

不得不说，露丝的偷盗行为对大熊猫造成了很大的伤害，影响十分恶劣。

It has to be said that Ruth’s theft has caused great harm to the giant panda, and the impact is very bad.

“六侠”赴欧美

Six Giant Pandas Went to West

1938年秋，英国人弗洛伊德·丹吉尔·史密斯带着6只大熊猫从成都启程前往英国。其中一只大熊猫受不了旅途艰辛死在途中。剩下的5只大熊猫于1938年12月24日抵达伦敦。史密斯给它们分别取名为“老奶奶”“开心果”“爱生气”“糊涂蛋”及“贝贝”。

In the autumn of 1938, Floyd Tangier Smith, left Chengdu with 6 giant pandas to England. One of the giant pandas could not stand the arduous journey and died on the way. The remaining five pandas arrived in London on December 24, 1938. Smith named them Grandma, Happy, Dopey, Grumpy and Baby.

养于成都华西协合大学的大熊猫
Giant pandas kept in West China Union University in Chengdu

“老奶奶”抵达英国仅仅两周就死于肺炎。自1939年1月26日起，“开心果”先后被带到德国柏林、汉诺威、慕尼黑、莱比锡和科隆动物园巡展。从1939年5月24日至1939年6月6日，“开心果”在法国巴黎的文森动物园展出。“开心果”于1939年6月6日从法国西北的瑟堡港乘坐远洋班轮赴美。1939年6月24日，“开心果”抵达美国圣路易斯动物园，直到1946年3月10日去世，终年11岁。期间，曾一度有另外一只大熊猫“宝贝”相伴。

Grandma, the eldest one, caught pneumonia and died only two weeks after arriving in Great Britain. Since January 26, 1939, Happy the giant panda was taken successively to Zoologischer Garten Berlin, Zoo Hannover, Münchner Tierpark Hellabrunn, Zoologischer Garten Leipzig & Kölner Zoo. After his German Tour, Happy was on display at the Zoo de Vincennes in Paris, France from May 24, 1939 until June 6, 1939. Happy left Europe on an ocean liner at the port of Cherbourg on June 6, 1939. He arrived in Saint Louis zoo, Saint Louis, USA on June 24, 1939 and stayed there for nearly 7 years, until his death on March 10, 1946. There once was another giant panda Pao-Pei accompanying him for some time.

其他3只大熊猫被安置在伦敦动物园并以中国朝代重新命名为“唐”“宋”“明”。“宋”于1939年12月18日因病去世，“唐”于1940年4月23日去世，最后只剩下大熊猫“明”。

The other three pandas were placed at the London Zoo and renamed as Tang, Sung and Ming for Chinese dynasties. Sung died on December 18, 1939, due to illness, Tang died on April 23, 1940, only the giant panda Ming was left.

聪“明”动英伦
Cute Ming Moved the UK

“明”是第一批来到英国的大熊猫之一。“明”的到来引起了人们的巨大兴趣，它成了伦敦动物园的焦点。人们蜂拥而至熊猫馆，只要能亲眼看看大熊猫的样子，无论多久他们都愿意等。在大熊猫“明”的众多粉丝中，英国著名摄影家伯特·哈迪和当时英国王室公主，后来的英国女王伊丽莎白和玛格丽特三位最有名。

Ming was one of the first giant pandas to came to Britain. It created massive interest and became the focus of the entire London Zoo. People flocked to the panda house and would wait as long as they could see pandas with their own eyes. Among all the fans of the giant panda, the British photographer Bert Hardy, one of Britain's best-known photographers, and Princess Elizabeth, who would become the queen of the United Kingdom of Great Britain and Northern Ireland, and her younger sister Princess Margaret, were the three most famous.

大熊猫摄影师“明”
Ming the photographer

1938年，伯特·哈迪带着儿子迈克到动物园为熊猫拍照。哈迪架好摄像机等待着最佳的拍摄时机，大熊猫“明”爬到相机前，像人一样站起来摆弄着摄像机，满脸的兴奋和好奇。哈迪搬过来一把椅子让迈克坐上去，并拍下了“明”为儿子拍照的照片。

In 1938, Bert Hardy took his son Mike on a special trip to the zoo to take pictures of pandas. Hardy set the camera for the best time to shoot. Ming climbed to the camera, stood up like a human to play with the camera with a full face of excitement and curiosity. Hardy brought in a chair for Mike to sit on, and captured a shot of a playful Ming behind one of his cameras on a tripod, seemingly taking a picture of his son, Mike.

1939年 3 月10日，“明”受到了玛丽王后、长公主玛丽及其丈夫拉塞尔斯子爵的“接见”。5月10日，伊丽莎白和玛格丽特来到伦敦动物园熊猫馆看“明”并进行了亲密接触。当时，英国正在抵抗德国法西斯的侵略。乐观开朗的“明”鼓舞了民众的士气，成了战时英雄和伦敦的名片。遗憾的是，“明”在第二次世界大战胜利前夕悄然去世了。

On March 10, 1939, Ming the giant panda “was received” by Queen Mary, the Grand Princess Mary and her husband viscount Lascelles. On May 10, Princess Elizabeth and her sister Margaret payed a visit to Ming and had close contact with it. At that time, Britain was resisting the aggression of Germany fascist. The optimistic Ming encouraged the morale of the people, and Ming became the wartime hero and the card of London. Unfortunately, Ming died quietly in 1944 on the eve of the victory of World War II.

为铭记“明”在促进中英人文交流方面的贡献，2015年10月21日，中华人民共和国国务院新闻办公室、中国日报社及中国人民对外友好协会联合向伦敦动物园赠送“明”的雕像。“明”以另一种方式回到伦敦，继续给伦敦人民带来欢乐和幸福。

伦敦动物园的大熊猫“明”的雕塑
Statue of Ming at the London Zoo
图片来源：sollyn

In order to remember the role of Ming the giant panda in promoting communication between China and the UK, The State Council Information Office of the People's Republic of China, the China Daily and the Chinese People's Association for Friendship with Foreign Countries jointly presented a statue of Ming to the London Zoo on October 21, 2015. Ming returned to London in another way and continued to bring joy and happiness to the people of London.

“潘多拉”联谊中美
Pandora Bridged China and American

1938年3月，华西协合大学的合作伙伴纽约动物协会向华西协合大学提出“希望得到一只大熊猫幼仔，如果可能的话最好是一对幼仔”。不久，华西协合大学生物系主任富兰克·丁克生的夫人从都江堰的大山中带回了一只活泼可爱的大熊猫幼仔，宠物般地养在华西坝家中，并取名为“潘多拉”。

In March 1938, West China Union University (WCUU) received a request from the New York Zoological Society, a partner of WCUU, to “get a giant panda baby, if possible, a pair of cubs”. Soon, Mrs. Dickinson, wife of Frank Dickinson, the dean of the Department of Biology, WCUU, brought back a lively and lovely giant panda cub, kept it as a pet, and named it Pandora.

“潘多拉”在华西坝生活了3个多月。1938年5月18日，华西协合大学化学系主任罗伊·斯普纳将其带到美国，落户纽约布朗克斯动物园。“潘多拉”成为第三只登陆美国的活体大熊猫。1939年，“潘多拉”被送到纽约世界博览会进行展览。在一年的展览期间，超过13万的美国人参观了“潘多拉”。1941年5月13日“潘多拉”去世。

Pandora the giant panda had lived in WCUU for more than three months. On May 18, 1938, Roy C. Spooner, dean of the Chemistry Department of WCUU, brought it to Bronx Zoo in New York City. Pandora became the third living giant panda landed in the United States. In 1939, Pandora was sent to the New York World's Fair for an exhibition. During the one-year exhibition, more than 130,000 Americans visited Pandora. Pandora died on May 13, 1941.

斯普纳与“潘多拉”在轮船上
Spooner with Pandora

淘气又可爱的“潘多拉”经常在华西协合大学的大草坪上玩耍，那些外籍教职人员及加拿大学校的孩子们经常一放学就来跟它亲密接触。“潘多拉”成了孩子们童年中最美好的记忆。80年后，当年的“CS孩子”（指那些在四川度过了他们少年时光的加拿大志愿者子女），世界上资格最老的大熊猫粉丝，曾再次回到成都，探寻儿时记忆，赓续中加友谊。“潘多拉”是中外文化交流的有力见证，是当之无愧的友好使者。

加拿大学校孩子与“潘多拉”
CS kids play with Pandora

The naughty and lovely Pandora often played on the big lawn of WCUU, and the foreign faculty and children in Canadian School (CS) always came to play with Pandora immediately after school. Pandora also became the best memory of the CS kids. Eighty years later, these CS kids, the oldest panda fans in the world, came back to Chengdu to explore their childhood memories and continue the friendship between China and Canada. Pandora became a strong witness of the cultural exchanges between China and the West and is a worthy envoy of friendship.

“三娇子”聚首美国

Three Giant Pandas Gathered in USA

1939年2月，美国商人戈登·坎贝尔把约1岁的雌性大熊猫“宝贝”带到美国，最后落脚于美国圣路易斯动物园。“宝贝”在美国引起了极大轰动，很长一段时间内，动物园天天游客如织，最高兴的人莫过于动物园园长亨利了，他对“宝贝”百般宠爱，天天与它待在一起。1939年6月24日，戈登从英国带来了大熊猫“开心果”，希望它能与“宝贝”配对产子。20多年后，“宝贝”走到了生命的尽头。

In February 1939, American businessman Gordon Campbell brought the one-year-old female giant panda Pao-Pei to the United States and finally settled at the St. Louis Zoo. Pao-Pei caused a great sensation in the United States. For a long time, the zoo was full of visitors every day. The happiest person was Henry, the director of St. Louis Zoo. He loved Pao-Pei in every way and stayed with her every day. On June 24, 1939, Gordon Campbell brought Happy the giant panda to company with Pao-Pei, wishing they could have their own child. Twenty years later, Pao-Pei died.

1939年，一位中国猎人抓到了“潘”并将它交给华西协合大学。1939年5月1日，迪安·赛奇夫人将它送到纽约动物园。“潘”不适应圈养环境，于1940年5月5日逝世。

In 1939, a Chinese hunter captured Pan the giant panda and handed it to West China Union University. On May 1, 1939, Mrs. Dean Sage escorted it to the New York Zoo. Pan couldn't adapt to the captive environment and died on May 5, 1940.

1939年秋，《芝加哥每日新闻》记者斯蒂尔得到一只大熊猫“美兰”。在摄影师罗伊·斯科特的护送下，“美兰”于11月16日到达芝加哥动物园。“美兰”体型健硕，在动物园悠闲地度过了一段时间，死于1953年9月5日，整整活了15年。

In the autumn of 1939, A. T. Steele, the *Chicago Daily News* reporter, got a giant panda named Mei-Lan. Escorted by photographer Roy Scott, Mei-Lan arrived at the Chicago Zoological Park on November 16. As a strong animal, Mei-Lan spent leisurely a period of time in the zoo, and died on September 5, 1953, living for 15 years.

斯科特与大熊猫“美兰”
Roy Scott with Mei-Lan

第二篇
至尊国礼

1941年，国民政府向美国援华救济联合会赠送大熊猫。中华人民共和国成立后，国家继续以政府和人民的名义将大熊猫作为国礼赠予那些与中国保持良好外交关系或中国希望与之建立外交关系的国家。先后共有美国、英国、苏联、朝鲜、日本、法国、墨西哥、西班牙及德国等9个国家接受过中国赠予的共计27只大熊猫。1982年，为响应保护濒危动物的全球性号召，中国政府宣布从1982年开始停止向外国赠送大熊猫。

Chapter Ⅱ State Gift

In 1941, the Kuomintang government presented giant pandas to the United China Relief. After the founding of the People's Republic of China, China continued to give pandas as national gifts in the name of the government and the people to those countries that maintained good diplomatic relations with China or that China wanted to establish diplomatic relations with. A total of 27 giant pandas had been donated by China to nine countries, including the United States, Britain, the Soviet Union, Democratic People's Republic of Korea, Japan, France, Mexico, Spain and Germany. In 1982, in response to the global call to protect endangered animals, the Chinese government announced that it would stop sending giant pandas abroad from 1982.

“潘弟”“潘达”谢美援助

Pan-Dee & Pan-Dah Appreciated American's Aid

1941年，中国人民抗日战争进入最艰苦的岁月。为感谢美国援华救济联合会的援助，宋美龄及其姐姐宋霭龄决定赠送美国一件具有中国特色的珍贵礼物。宋氏姐妹认为大熊猫就是最佳礼物。

In 1941, the Chinese People's War of Resistance against Japanese Aggression entered the hardest years. In order to thank the United China Relief (UCR) for its assistance, Soong Mei-ling and her sister Soong Ai-ling, decided to give a precious gift with Chinese characteristics to the U.S. The Soong sisters thought the giant panda the best gift.

得知中国将赠送熊猫的消息，美方十分兴奋，立即派纽约动物协会会长约翰·蒂文专程来华接收。10月30日，蒂文飞抵成都，11月6日，他护送着两只熊猫由成都飞抵重庆，准备参加重庆国民政府举行的大熊猫赠送仪式。

The U.S. was very excited to learn of the news that China would give the giant panda as a present, and immediately sent John Tee-Van, president of the New York Zoological Society, to China to accept the pandas. On October 30, Tee-Van flew to Chengdu. On November 6, he conveyed two pandas from Chengdu to Chongqing to attend the ceremony of panda-giving held by the Kuomintang government in Chongqing.

葛维汉与约翰・蒂文与大熊猫嬉戏
Graham & John Tee-Van play with the giant panda

1941年11月9日，大熊猫赠送仪式在重庆举行，美国驻华大使高思，美方代表蒂文和成都华西协合大学博物馆馆长葛维汉教授等美方代表及国民政府官员出席典礼。典礼上，宋氏姐妹代表国民政府正式宣布将大熊猫作为国礼赠送给美国。

On November 9, 1941, the ceremony of giving giant pandas to the UCR was officially held in Chongqing. Clarence E. Gauss, American Ambassador to China, John Tee-Van, American representative, and Professor David Crockett Graham, director of the Museum of West China Union University in Chengdu, and the officials of the Kuomintang government attended the ceremony. At the ceremony, the Soong sisters announced on behalf of the Kuomintang government that they would give pandas to the United States as a national gift.

11月14日，两只大熊猫从重庆出发并于12月25日抵达旧金山，12月30日抵达纽约，随后被安置在纽约市布朗克斯动物园。1942年4月29日，两只大熊猫正式取名为“潘弟”（“班棣”）和“潘达”（“班达”）。

On November 14, two giant pandas departed from Chongqing and arrived in San Francisco on December 25, and New York on December 30, and was then placed at the Bronx Zoo in New York City. On April 29, 1942, the two giant pandas were officially named Pan-Dee and Pan-Dah.

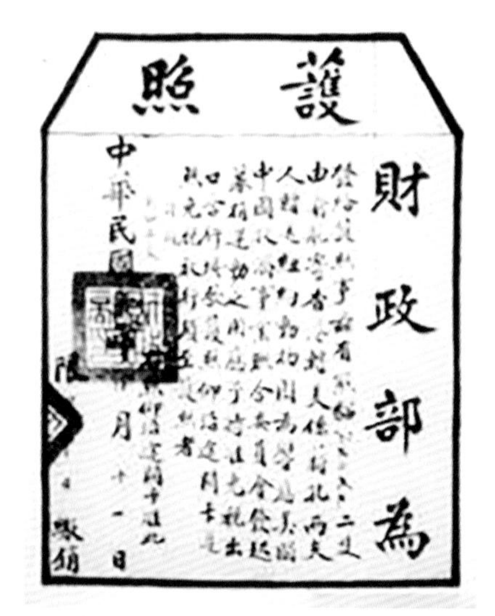

護照

財政部為

中華民國

月 十一 日

“潘弟”和“潘达”及其出国护照
Pan-Dee & Pan-Dah and their passport

大熊猫赴美途中，恰逢珍珠港事件爆发。战争带来的痛苦与熊猫带来的欢乐，在美国人民心中形成了鲜明对比。两只熊猫在美国引起了广泛关注并深受喜爱，中美两国之间的友谊也因大熊猫而更加深厚。“潘弟”于1945年10月4日死于腹膜炎。“潘达”在布朗克斯动物园生活了近10年，于1951年10月31日去世。

The giant panda’s way to the United States coincided with the outbreak of the Pearl Harbor incident. The pain of the war was in sharp contrast to the joy of the pandas in the hearts of the American people. The two pandas had attracted wide attention and love in the United States, and the friendship between China and the United States has been deepened by the pandas. Pan-Dee died of peritonitis on October 4, 1945, and Pan-Dah on October 31, 1951, living at the Bronx Zoo for nearly 10 years.

“兰兰”“康康”首渡日本
Lan-Lan & Kang-Kang Landed in Japan

1972年9月25日，日本首相田中角荣访华。9月29日，中日双方签署联合声明，正式建立外交关系。为纪念中日邦交正常化，在日本首相的一再恳求下，中国同意将大熊猫“兰兰”和“康康”作为“礼物”赠送给日本。

On September 25, 1972, Kakuei Tanaka, then Prime Minister of Japan, made a visit to China. On September 29, China and Japan signed the joint statement to establish diplomatic relations. After repeated requests from the Japanese prime minister, China agreed to give Japanese a pair of giant pandas to commemorate the normalization of Sino-Japanese diplomatic relations. Giant pandas Lan-Lan and Kang-Kang were presented to the Japanese.

1972年10月28日，大熊猫“兰兰”和“康康”抵日
Lan-Lan and Kang-Kang arrived in Japan on Oct. 28, 1972
图片来源：东京动物园协会

1972年10月28日，当两只大熊猫抵达日本时，日本内阁官房长官二阶堂进亲自迎接。11月4日，上野公园举行隆重的仪式欢迎大熊猫定居日本。日方负责人在仪式上称："这一对熊猫是中国人民赠送给日本国民的最好礼物，11月4日是日中两国人民友好的象征。"11月5日，大熊猫正式露面。

On October 28, 1972, when two cute giant pandas flew into Japan, Chief officer Nikaid ō Susumu personally welcomed the pandas. On November 4, Ueno Park held a grand ceremony to welcome the pandas to settle down in Japan. At the ceremony, the Japanese officially said, "This pair of pandas are the best gift from the representative Chinese people to the Japanese people, and November 4 is a symbol of the friendship between the Japanese and Chinese people." On November 5, the giant panda officially appeared in front of the Japanese people.

1972年11月5日，日本民众排队参观“兰兰”和“康康”
The Japanese people visit Lan-Lan & Kang-Kang in line on Nov. 5, 1972

大熊猫的到来，在日本迅速掀起了一股“大熊猫热”，并引发了日本人的“中国热”。1979年9月4日，10岁的“兰兰”因急性肾功能不全并发尿毒症死亡，“康康”死于1980年6月30日。后来它们被制成标本存放在多摩动物园。

The arrival of giant pandas quickly set off a “panda-mania” in Japan, and also promoted the “Chinese-mania” among the Japanese people. On September 4, 1979, Lan-Lan died of acute kidney insufficiency complicated with uremia at the age of 10, and Kang-Kang died on June 30, 1980. They were made as specimens and stored in Tama Zoological Park.

雅安市雨城大桥上的“兰兰”雕塑
Statue of Lan-Lan on the Yucheng Bridge, Ya’an City
李守林　摄

雅安市雨城大桥上的“康康”雕塑
Statue of Kang-Kang on the Yucheng Bridge, Ya’an City
李守林　摄

“联合”抵达不列颠

Lien–Ho Arrived in Britain

1945年，为纪念抗日战争的胜利，在英国政府的请求下，国民政府决定赠送一只大熊猫给英国。1946年5月5日，在四川大学生物系教师马德（马骥群）的护送下，年仅1岁的大熊猫“联合”踏上了赴英之路。

In 1945, in order to commemorate the victory of the War of Resistance Against Japanese Aggression, at the request, the Kuomintang government decided to send a giant panda to the Britain. On May 5,1946, under the escort of Ma Chi-chun, Professor of Department of Biology, Sichuan University, the 1-year-old giant panda Lien-Ho (Unity) set foot on the road to Britain.

马骥群与大熊猫“联合”
Ma Chi-chun takes photo with Lien-Ho
图片来源：四川大学档案馆

5月6日“联合”乘坐专机经昆明，到加尔各答，最后于5月11日到达伦敦摄政公园动物园。它的到来在当地引起了轰动。无数电话询问开展时间，各大报刊也刊登它的照片、漫画，以及与它相关的各种消息、论文，所占篇幅超过了同时期任何政要、名人。“联合”于1950年2月22日去世。

On May 6, Lien-Ho took a special plane from Kunming to Kolkata and finally arrived at the Regent's Park in London on May 11. Her arrival had caused a sensation among the British people. Countless phone inquired about the time of exhibition of the giant panda. The major newspapers also published her photos, cartoons, messages and papers, taking more space than any politician or celebrity of the same time. Lien-Ho died on February 22, 1950.

“平平”“安安”迁居莫斯科
Ping-Ping & An-An Moved to Moscow

中华人民共和国成立后，大熊猫开始作为外交使者在中国的对外交往中扮演积极角色。1957年4月至5月，俄国十月革命胜利40周年之际，苏联最高苏维埃主席团主席伏罗希洛夫访华。在参观北京动物园时，他被憨态可掬的大熊猫吸引，便请求中国政府赠送一对大熊猫给苏联。

After the founding of the People's Republic of China, giant pandas began to play a positive role in China's foreign exchanges as diplomatic emissary. From April to May, 1957, the Soviet Union Chairman Kliment Voroshilov visited China. While visiting the Beijing Zoo, he was attracted by the naive pandas and asked the Chinese government to give a pair of giant pandas to the Soviet Union.

1957年，莫斯科动物园中的大熊猫“平平”
Ping-Ping in Moscow State Zoo, 1957
图片来源：《大熊猫史话1869—2019》

1957年5月18日，雄性大熊猫“平平”以“国礼”形式送给莫斯科国家动物园。它是中华人民共和国成立后首次走出国门的“国礼”大熊猫。1959年8月18日，雄性大熊猫“安安”（被误认为雌性）被送给莫斯科。1961年，“平平”去世，1974年“安安”去世。

On May 18, 1957, Ping-Ping the male giant panda was presented to the Soviet Union as a “State Gift” and lived in Moscow State Zoo. Ping-Ping was the first panda going abroad after the founding of the People’s Republic of China. And An-An, another male giant panda (mistaken as female) was sent to Moscow on August 18, 1959. Ping-Ping died in 1961, and An-An died in 1974.

雅安市雨城大桥上的“平平”“安安”雕塑
Statue of Ping-Ping & An-An on the Yucheng Bridge, Ya’an City
李守林 摄

五熊猫奔赴朝鲜
Five Pandas Went to DPRK

1959年7月，为表达友谊，在朝鲜中央动物园建成之际，中国向朝鲜赠送了一批珍禽异兽。1965年6月3日，中国向朝鲜赠送了一对大熊猫“一号”和“二号”。朝鲜成为继苏联之后，第二个获得中国赠送大熊猫的国家。

In July 1959, when the Central Zoo of Democratic People’s Republic of Korea(DPRK) was built, China presented a number of rare birds and animals to DPRK to express friendship. On June 3, 1965, China gifted a couple of giant pandas No. 1 and No. 2 to DPRK. Therefore, DPRK became the second country to receive giant pandas after the Soviet Union.

雅安市雨城大桥上的“一号”“二号”雕塑
Statue of No. 1 & No. 2 on the Yucheng Bridge, Ya’an City
李守林　摄

1971年10月20日，中国再次送给朝鲜一对大熊猫——雄性“凌凌”和雌性“三星”。“凌凌”于1972年10 月18日去世。1979年3月20日，中国又赠送了一只雄性大熊猫“丹丹”。多次赠予朝鲜大熊猫，反映了中国对于中朝友谊的重视。

On Oct. 20, 1971, China gave a pair of giant pandas, the male one Ling-Ling and the female one San-Xing, to DPRK, and Ling-Ling died on October 18, 1972. On March 20, 1979, China gave a male giant panda Dan-Dan to DPRK. The repeated gifts to DPRK reflects the importance China attaches to China-DPRK friendship.

大熊猫“丹丹”
Dan-Dan the Giant Panda
图片来源：《大熊猫史话1869—2019》

雅安市雨城大桥上的“丹丹”雕塑
Statue of Dan-Dan on the Yucheng Bridge, Ya’an City
李守林　摄

1991年朝鲜发行的大熊猫邮票
Giant panda stamps issued in Democratic People’s Republic of Korea, 1991

传奇“姬姬”扬名四海

Chi-Chi Known Around the World

1958年，美国芝加哥动物协会告诉奥地利动物商人海尼·德默，如果他能从中国得到一只大熊猫，他们愿意以25 000美元的价格购买。随后，海尼·德默以3只长颈鹿、2只犀牛、2只河马及2只斑马等动物与北京动物园交换得到大熊猫“碛碛”。但美国政府出于政治原因拒绝大熊猫入境，海尼·德默只得将“碛碛”带到欧洲伺机售卖。

In 1958, the American Zoological Society of Chicago told Austrian animal dealer Heini Demmer that they would pay $25,000 for the giant panda if he could get one from China. Later, Heini Demmer got a giant panda Qi-Qi from Beijing Zoo exchanging with three giraffes, two rhinos, two hippos and two zebras. But the US government refused the Chinese panda to enter for political reason, so Heini Demmer had to take Qi-Qi to Europe to look for an opportunity to sale.

1958年5月，“碛碛”抵达莫斯科。随后，“碛碛”先后短暂停留德国法兰克福动物园、丹麦哥本哈根动物园以及德国东柏林动物园，最后伦敦动物园以1.2万英镑买下“碛碛”并改名“姬姬”。9月26日，“姬姬”正式入住伦敦动物园，结束了颠沛流离的生活。“姬姬”也成为中国历史上唯一一只以交换形式送往外国的大熊猫。

In May 1958, Qi-Qi arrived in Moscow. Qi-Qi visited the Frankfurt Zoo, the Copenhagen Zoo in Denmark and the Tiergarten Berlin in East Germany. Finally, the London Zoo bought Qi-Qi for 12,000 and renamed it Chi-Chi. On September 26, Chi-Chi came to London and ended her vagabondage life. Chi-Chi also became the only giant panda in Chinese history to be sent to foreign countries in the form of exchange.

雅安市雨城大桥上的“姬姬”雕塑
Statue of Chi-Chi on the Yucheng Bridge, Ya’an City
李守林　摄

有趣的是，为与大熊猫“安安”联姻，“姬姬”还曾于1966年3月11日前往莫斯科并于10月17返回伦敦。1968年8月，“安安”也曾前往伦敦动物园与“姬姬”共同生活9个月。作为第二次世界大战后第一只抵达欧洲的大熊猫，“姬姬”所到之处无不受到民众的热烈欢迎。遗憾的是，直到1972年7月22日病逝，“姬姬”也没有生下自己的孩子。“姬姬”去世后受到了公开哀悼。1973年“姬姬”被制作成标本并存放在英国自然历史博物馆中。

Interestingly, Chi-Chi had been to Moscow in her marriage to An-An the giant panda on March 11, 1966, and returned to London on October 17. In August 1968, An-An also went to the London Zoo to live with Chi-Chi for nine months. As the first giant panda arrived in Europe after World War II, Chi-Chi was warmly welcomed by the public everywhere she went. Unfortunately, until her death on July 22, 1972, at the age of 18, Chi-Chi did not have birth of her own. Chi-Chi was openly mourned after her death. In 1973, Chi-Chi was made into specimen and stored in the British Museum of Natural History.

1961年，世界野生生物基金会刚成立之时，希望找到一个能代表地球珍贵濒危物种且被广泛认可的动物作为基金会的标志。基金会创始人之一、英国自然历史学家和画家彼得·斯科特爵士以大熊猫“姬姬”为原型手绘的大熊猫图案被采纳。大熊猫“姬姬”的形象还成了世界野生生物基金会会旗、会徽上的标志，其被誉为“全世界最有名的熊猫”。

In 1961, the World Wildlife Fund (WWF) was first founded and they wanted to find a widely accepted symbol which could represent the rare and endangered species on earth. Sir Peter Scott, one of WWF's founders and an English natural historian and painter, drew a pattern of Chi-Chi the giant panda and it then was adopted as the symbol of WWF. The image of Chi-Chi has also become the symbol on the flag and emblem of the WWF, and Chi-Chi has been known as “the most famous panda in the world”.

不同时代的WWF标志
Logos of WWF in different eras
图片来源：《大熊猫文化笔记》

"玲玲""兴兴"牵手美利坚

Ling-Ling & Hsing-Hsing Bridged USA and China

1972年2月21日，时任美国总统的尼克松携夫人对中国进行了历史性访问。尼克松夫人帕特·尼克松抵达北京后的第二天即赴北京动物园参观并同大熊猫进行了亲密接触。在27日的告别晚宴上，中华人民共和国总理周恩来表示中国将赠送2只大熊猫给美国。

On February 21, 1972, then US President Nixon and his wife paid a historic visit to China. Mrs. Pat Nixon went to the Beijing Zoo to visit pandas the second day they arrived and had close contact with giant pandas. At the farewell dinner on February 27, before they left China, the Chinese Premier Zhou Enlai announced that China would present the US with two pandas.

赠送大熊猫给美国是纪念中美两国领导人开创性握手的最好方式。4月16日，大熊猫“玲玲”和“兴兴”乘坐专机飞抵马里兰州的安德鲁斯空军基地并由警察护送到华盛顿国家动物园。它们受到非常高级别的礼遇，第一夫人帕特·尼克松亲自主持官方接收仪式，1000多位政要名人出席欢迎仪式，8000多名民众冒雨相迎。大熊猫的到来掀开了中美历史新篇章。1972年在美国也被称为“熊猫年”。

Giving giant pandas to the US was the best way to commemorate the groundbreaking handshakes between leaders from China and USA. On April 16, giant pandas Ling-Ling and Hsing-Hsing arrived at the Andrews Air Force Base in Maryland and were taken by police escort to the Smithsonian's National Zoo in Washington, DC. They received very high-level hospitality. The first lady Pat Nixon personally presided over the official reception ceremony. More than 1,000 dignitaries attended the welcoming ceremony, and over 8,000 Americans greeted Ling-Ling and Hsing-Hsing on their arrival. The arrival of the pandas marked a new chapter in the US-China history. And the year of 1972 was marked as the “Year of the Panda” in the United States.

1992年12月31日，大熊猫“玲玲”走完23年生活历程，无疾而终，大熊猫“兴兴”于1999年11月28日因肾衰竭被施行安乐死。作为中美友谊的重要象征，“玲玲”和“兴兴”不仅引起了美国各地的熊猫热，也引起了美国人民对中国的兴趣。50多年来，大熊猫一直是两国之间友谊的象征。2022年3月16日，美国国家动物园拉开了为期半年的大熊猫抵美50周年系列庆祝活动序幕，以纪念中美两国在大熊猫交流和保护方面的紧密合作。

On December 31, 1992, the giant panda Ling-Ling finished 23 years of life, while Hsing-Hsing was euthanized on November 28, 1999 due to kidney failure. Served as an important symbol of friendship between the two countries, Ling-Ling and Hsing-Hsing aroused not only Panda-mania across the country, but also interest in China among the American people. For five decades since then, giant pandas have remained a symbol of friendship between the Chinese and American people. On March 16, 2022, the National Zoo of the United States began a half-year celebration of the 50th Pandaversary of the arrival in the United States to commemorate the close cooperation between China and the United States in the exchange and protection of the giant panda.

雅安市雨城大桥上的“玲玲”“兴兴”雕塑
Statue of Ling-Ling & Hsing-Hsing on the Yucheng Bridge, Ya’an City
李守林　摄

“燕燕”“黎黎”情定法兰西

Yen-Yen & Li-Li Kissed in France

1973年9月11日，法国总统乔治·让·蓬皮杜访问中国。他是1964年中法建交以来第一个访华的在任西欧国家元首。在蓬皮杜的请求下，大熊猫“黎黎”和“燕燕”被中国政府送给法国。

On September 11, 1973, French President Georges Jean Pompidou visited China. He was the first sitting head of state in Western European countries to visit China since the establishment of diplomatic ties between China and France in 1964. At the request of the Pompidou, giant pandas Li-Li and Yen-Yen were sent to France by Chinese government.

雅安市雨城大桥上的“黎黎”雕塑
Statue of Li-Li on the Yucheng Bridge, Ya’an City
李守林　摄

1973年12月8日，“燕燕”和“黎黎”抵达巴黎并定居于巴黎文森动物园。他们的到来在法国卷起一股熊猫热，蓬皮杜总统也因此威望倍增。1981年，法国总统德斯坦赴动物园参观大熊猫，并进入“燕燕”的卧室与其合影留念。“燕燕”立即像人一样站起来并伸出前掌朝总统走去，像是要拥抱总统一样。

On December 8, 1973, Yen-Yen and Li-Li arrived in Paris and settled at the Vincennes Zoo. Their arrival rolled up a Panda-mania in France, and the President Georges Pompidou had also increased his prestige for that. In 1981, the president Valery Giscard d’Estaing visited Vincennes Zoo and entered the bedroom of Yen-Yen to take a photo with it. And Yen-Yen stood up immediately with his back feet like humans and stretched out his front feet as if he wanted to embrace the president.

“黎黎”于1974年4月20日辞世后，“燕燕”成了该动物园最大的明星。它活了27岁，于2000年1月20日去世。就像多年前的“黎黎”一样，“燕燕”也是死于胰腺肿瘤。自此一直到2012年，法国人民再也没有在法国见过大熊猫。“燕燕”和“黎黎”赴法在中法友谊史上具有里程碑意义。

Li-Li died on April 20, 1974, and Yen-Yen became the biggest star of the zoo until his death on January 20, 2000. Just like Li-Li, many years before, the cause of his death was a pancreas tumour. He became 27 years old. Since then, the French people had never seen a giant panda again until 2012. Yen-Yen and Li-Li living in France is a milestone in the history of China-France friendship.

“佳佳”“晶晶”再赴英格兰
Chia-Chia & Ching-Ching Reentered the UK

大熊猫“姬姬”的过世让英国民众的生活出现了一个熊猫空洞，每个人都想把这个遗憾弥补起来。1974年5月24日，渴望加强与中国的关系的英国首相爱德华·希思访问中国并代表英国请求中国赠送两只大熊猫。1974年9月13日，大熊猫“佳佳”和“晶晶”搭乘专机直飞伦敦。

The death of Chi-Chi has made the British feel hollow inside, and everyone wanted to make up for it. In May 24, 1974, Edward Heath, the former Prime Minister of UK, keen to foster relations with China, visited China and asked on behalf of his country for two giant pandas. On September 13, 1974, Chia-Chia and Ching-Ching took a flight directly to London.

“佳佳”和“晶晶”的首次露面就吸引了近两万名观众，尽管当天下着瓢泼大雨，民众的热情却丝毫不受影响。希思也冒雨看望大熊猫，并在参观之后兴高采烈地对记者发表了讲话。他说到，大熊猫的到来，显示了中国政府对英国人民的情谊，它们必将受到英国人民的热烈欢迎。新的熊猫和新的外交格局预示着中英关系走上了新的里程。

The first appearance of Chia-Chia and Ching-Ching attracted nearly 20,000 viewers. Despite the pouring rain, the enthusiasm of the public was not affected. Former Prime Minister Edward Heath also visited the pandas in the rain and spoke cheerfully to reporters after the visit. He said that the arrival of the pandas showed the friendship of the Chinese government to the British people, and the pandas would be warmly welcomed by the British people. New panda and new diplomatic landscape herald a new milestone in China-UK relations.

雅安市雨城大桥上的“佳佳”雕塑
Statue of Chia-Chia on the Yucheng Bridge, Ya’an City
李守林　摄

“晶晶”身体不好，经常需要进行医学观察。1985年，“晶晶”亡故，没有生育。而“佳佳”拥有良好的繁殖记录，先后前往美国华盛顿国家动物园、辛辛那提动物园以及墨西哥查普特佩克动物园配种。1991年10月13日，大熊猫“佳佳”在墨西哥查普特佩克动物园去世，结束了其17年的“熊猫特使”生涯。

Ching-Ching was in poor health and often needed medical observation. In 1985, Ching-Ching died without offspring. Chia-Chia, having a good breeding record, went to mate successively to the Washington National Zoo, Cincinnati Zoo and Chapultepec Zoo, where he lived to die on October 13, 1991 and ended his 17-year career as a panda envoy.

“迎迎”“贝贝”繁衍墨西哥
Ying-Ying & Pe-Pe Left for Mexico

1972年2月14日，中国与墨西哥正式建立外交关系。在墨西哥总统路易斯·埃切维里亚·阿尔瓦雷斯的请求下，中国同意赠送一对大熊猫给墨西哥，以庆贺中墨建交三周年。当年9月10日，雌性大熊猫“迎迎”和雄性大熊猫“贝贝”启程前往墨西哥，并在查普尔特佩克动物园安家，这也是拉丁美洲唯一拥有大熊猫的动物园。

On February 14, 1972, China and Mexico formally established diplomatic relations. At the request of Mexican President Luis Echeverría Álvarez, China agreed to give a couple of pandas to Mexico to celebrate the third anniversary of the establishment of diplomatic ties between China and Mexico. On September 10, Ying-Ying, a female panda, and Pe-Pe, a male one, left for Mexico and made their home at the Chapultepec Zoo, which is the only zoo in Latin America with giant pandas.

雅安市雨城大桥上的“迎迎”雕塑
Statue of Ying-Ying on the Yucheng Bridge, Ya’an City
李守林　摄

对不产竹子的墨西哥来说，饲养大熊猫并不是一件简单的事情。饲养人员和科研人员尝试将可食用的仙人掌喂养大熊猫。令人惊讶的是，富含多种维生素和微量元素的仙人掌成了大熊猫的最爱。良好的生活条件及饲养人员和科研人员的细心照顾，令“贝贝”和“迎迎”成为中国以外最高产的大熊猫夫妻，它们共孕育了7个孩子。

In Mexico, raising giant pandas is not an easy task. Zoo keepers and researchers tried hard to feed giant pandas with the eatable cactus. It’s surprising that cactus, rich in various vitamin and trace elements, became the giant pandas’ favourite food. Good living conditions and the great care of the keepers and researchers made Pe-Pe and Ying-Ying the highest producing pandas outside of China, breeding seven kids.

大熊猫在墨西哥深受欢迎，墨西哥不仅有家喻户晓的大熊猫主题歌曲《查普特佩克的大熊猫》，还有大熊猫纪念币。1988年7月20日，“贝贝”死于癌症，1989年1月29日，“迎迎”去世。为纪念“迎迎”和“贝贝”，查普特佩克动物园为他们塑造了雕像。出生于1990年7月1日的第三代“欣欣”是墨西哥现存唯一一只大熊猫，也是目前拉丁美洲唯一的大熊猫。“迎贝”家族是中国和墨西哥两国友谊的象征。

Giant pandas was well received in Mexico. Mexico has not only the well-known giant panda theme song “*El pequeño panda de Chapultepec*”, but also the giant panda commemorative coins. Pe-Pe died of cancer on July 20, 1988, and Ying-Ying died on January 29, 1989. Pe-Pe and Ying-Ying are a heroic couple. In honor of Ying-Ying and Pe-Pe, Chapultepec Zoo created statues of them. Xin-Xin, born on July 1, 1990, their third generation, is the only giant panda in Mexico. It is currently the only giant panda in Latin America, as well. The Ying-Pe family is a symbol of a unique friendship between China and Mexico.

“强强”“绍绍”舞动西班牙

Chang-Chang & Shao-Shao Danced in Spain

1978年6月17日，西班牙国王胡安·卡洛斯一世和索菲亚王后首次对中国进行国事访问。为表达友谊，中国赠送西班牙一对大熊猫。12月28日，大熊猫“强强”和“绍绍”飞抵马德里，成为首批抵达西班牙的中国大熊猫。

On June 17, 1978, King Juan Carlos I and Queen Sofia of Spain made their first state visit to China. The Chinese government presented a pair of giant pandas to the Spanish people to express the friendship. On December 28, Chang-Chang and Shao-Shao the giant pandas flew to Madrid and were the first Chinese giant pandas to arrive in Spain.

雅安市雨城大桥上的“绍绍”雕塑
Statue of Shao-Shao on the Yucheng Bridge, Ya’an City
李守林　摄

1982年9月4日晚，“绍绍”分娩一对龙凤胎！这个消息让睡得正香的国王胡安·卡洛斯乐不可支并很快就传遍全世界。遗憾的是，只有一只雄性幼崽存活了下来，其成为首只在欧洲圈养环境下出生并存活的大熊猫。西班牙在全国为熊猫幼崽征求名字，最后一致同意取名“竹琳”，寓意“竹林之宝”。

On the evening of September 4, 1982, Shao-Shao gave birth to a twins. The news made King Juan Carlos very happy when he was sleeping soundly and was spread all over the world. It’s a pity that only one male cub survived, becoming the first giant panda to be born and survive in European captivity. Spain asked for a name for the panda cub nationally, and finally all agreed the name Chu-Lin which means “the treasure of the bamboo forest”.

“欢欢”“飞飞”“陵陵”续缘日本

Huan-Huan & Fei-Fei & Ling-Ling Related Japan

1979年12月5日，日本首相大平正芳访华，中华人民共和国国务院总理华国锋代表中国向日本赠送大熊猫。1980年1月29日晚，“欢欢”抵达日本上野动物园，与“康康”作伴。

On December 5, 1979, when Japanese Prime Minister Masayoshi Ohira visited China, the Chinese Premier Hua Guofeng presented a giant panda to Japan. On the evening of January 29, 1980, Huan-Huan the female giant panda arrived at the Ueno Zoo in Japan, accompanying Kang-Kang.

为纪念中日邦交正常化十周年，1982年5月31日，中国表示将再次赠送一只大熊猫给日本。10月9日，大熊猫“飞飞”飞抵日本。“飞飞”的到来使上野动物园再次掀起大熊猫热，众多日本民众争相来动物园一睹“飞飞”的风采。

On May 31, 1982, China announced that would give another giant panda to Japan to commemorate the 10th anniversary of the normalization of diplomatic relations between China and Japan. On October 9, Fei-Fei flew to Japan. The arrival of Fei-Fei kept the Panda-mania up, and many Japanese rushed to the zoo to see the elegant demeanour of Fei-Fei.

雅安市雨城大桥上的“欢欢”雕塑
Statue of Huan-Huan on the Yucheng Bridge, Ya’an City
李守林　摄

“飞飞”是中国政府送给外国的最后一只大熊猫。“飞飞”和“欢欢”于1986年6月1日生下女儿“童童”，1988年6月23日，它们的儿子“悠悠”降生。1992年，中日邦交正常化20周年之际，日本明仁天皇访华，中国同意用“陵陵”与“悠悠”进行交换，以免近亲繁殖。“飞飞”“欢欢”和“童童”分别于1994年、1997年和2000年先后去世。2008年4月30日凌晨，“陵陵”因慢性心脏衰竭去世。

Fei-Fei is the last giant panda sent overseas as present by the Chinese government. Fei-Fei and Huan-Huan gave birth to their daughter Tong-Tong on June 1, 1986, and their son You-You on June 23, 1988. In 1992, on the 20th anniversary of the establishment of diplomatic ties between China and Japan, Japanese Emperor Akihito visited China, and China agreed to exchange Ling-Ling for You-You to avoid inbreeding. Fei-Fei, Huan-Huan and Tong-Tong died respectively in 1994, 1997 and 2000. In the early hours of April 30, 2008, Ling-Ling died of chronic heart failure.

"天天""宝宝"到达德国

Tian-Tian & Bao-Bao Went to Germany

1979年10月，中华人民共和国国务院总理华国锋访德时，德国总理赫尔穆特·施密特提出希望中国赠送大熊猫的请求。1980年，中国政府同意送给德国人民一对大熊猫。11月5日，大熊猫"宝宝"和"天天"抵达德国的柏林动物园，受到热情接待并成为该园的"镇园之宝"。

In October 1979, Chinese Premier Hua Guofeng visited Germany. Germany Chancellor Helmut Schmidt put forward the hope that China presents giant pandas to Germany. The Chinese government agreed. On November 5, 1980, Bao-Bao and Tian-Tian the giant pandas arrived at the Zoo Berlin in Germany and were warmly received and became the treasure of the zoo since then.

施密特总理一直关心着大熊猫。1980年11月8日，施密特总理和夫人来到柏林动物园看望大熊猫。“宝宝”和“天天”表现得特别热情、活泼，施密特夫妇被逗得哈哈大笑。施密特曾说道：“过去，熊是柏林城的象征，今后应该以熊猫为象征。”由此可见西德人民对大熊猫的喜爱程度。

Schmidt had always been concerned about the giant pandas. On November 8, 1980, Schmidt and his wife visited the pandas at the Zoo Berlin. Bao-Bao and Tian-Tian were particularly enthusiastic and lively, and the Schmidt were amused with laughter. Schmidt once said: “In the past, the bear was the symbol of Berlin, and in the future, it should be the panda.” This showed how the West German love for giant pandas.

雅安市雨城大桥上的“宝宝”雕塑
Statue of Bao-Bao on the Yucheng Bridge, Ya’an City
李守林　摄

雅安市雨城大桥上的“天天”雕塑
Statue of Tian-Tian on the Yucheng Bridge, Ya’an City
李守林　摄

遗憾的是，“天天”于1984年2月5日不幸去世。“宝宝”是幸运的，一直生活在德国人民的万千宠爱中。“宝宝”于2012年8月22日离世，活到了34岁，相当于人类102岁。至此，所有中国赠送出去的“国礼”大熊猫，全部已成往事。

It is a pity, Tian-Tian died on February 5, 1984. Bao-Bao was lucky and had been living in the favor of the German people. Bao-Bao lived to the ripe old age of 34, the equivalent of 102 in human terms, dying on the morning of August 22, 2012. So far, all the national gifts, giant pandas, that China gave out had become the past.

第三篇
商务参赞

据不完全统计，自1984年至1992年，先后有30多只大熊猫出国短期巡展，为增进我国人民同世界各国人民的文化交流和友好交往做出了贡献。鉴于大熊猫生态环境恶化，导致其数量急剧减少，为响应保护濒危动物的全球倡议，1992年，中国宣布停止租借大熊猫，但仍有大熊猫国际交流与合作的情况，在极个别情况下，仍有大熊猫短期出国的。由于资料缺失，本篇仅列举部分影响较大的大熊猫出国巡展事例。

Chapter Ⅲ Commercial Counsellor

According to the incomplete statistics, from 1984 to 1992, more than 30 giant pandas went abroad for a short-term display, which made contributions to enhancing the cultural and friendly exchanges between the Chinese people and the people of the world. In response to the global initiative to protect endangered animals, China announced in 1992 that it would stop leasing giant pandas for commercial purposes. But the door for international exchange and cooperation of pandas was not closed. In very rare cases, there were still some short-term overseas trips. For the lack of data, this part only lists some examples of the influential giant pandas displaying abroad.

“迎新”“永永”助兴奥运

Ying–Xin & Yong–Yong Pleased the Olympic Games

1984年，时任西方石油公司总裁、著名实业家哈默访华。期间，他向邓小平提出希望能向中国借两只熊猫为即将召开的洛杉矶奥运会助兴。鉴于1984年洛杉矶奥运会是中国恢复联合国席位和奥委会席位后首次参加的奥运会，中国同意短期租借一对熊猫给美国，以表示对洛杉矶奥运会的支持。

In 1984, Armand Hammer, then the president of Occidental Petroleum Corporation and a famous industrialist, visited China. During the meantime, he asked Deng Xiaoping, then the leader of China, to lend two giant pandas to please the upcoming Los Angeles Olympics. Considering it’s the first time China participating in the Olympics after restoring the seat in the UN and IOC, China agreed to lease a pair of pandas to the United States for a short term to show support for the Los Angeles Olympics.

7月13号，大熊猫“永永”和“迎新”飞抵洛杉矶，成为第一对以巡展方式走出国门的大熊猫。洛杉矶市政府官员、美国西方石油公司总裁哈默、洛杉矶与广州友好协会主席阿曼森夫人，以及一大群新闻记者到机场迎接大熊猫。1984年7月20日，“永永”和“迎新”正式与游客见面。奥运会结束后，它们又被租借给旧金山动物园进行了三个月的商业展出。后又相继赴加拿大、爱尔兰、瑞典、比利时等国巡展。“永永”和“迎新”助兴奥运揭开了大熊猫国际交流的新篇章。

On July 13th, two giant pandas Yong-Yong and Ying-Xin flew to Los Angeles, becoming the first pair of giant pandas to go abroad in the way of itinerant exhibition. Los Angeles city officials, Armand Hammer, the president of Occidental Petroleum Corporation, and Mrs. Amanson, the president of the Los Angeles and Guangzhou Friendship Association, and a large group of journalists came to the airport to greet the pandas. On July 20, 1984, the giant panda officially met with visitors. After the Olympic games, they were loaned to the San Francisco Zoo for three months of commercial display, and also in Canada, Ireland, Sweden, Belgium and other countries. Yong-Yong and Ying-Xin opened a new chapter in the international exchanges for giant pandas.

“平平”“明明”巡展爱尔兰

Ping-Ping & Ming-Ming Exhibited in Ireland

1986年6月11日，大熊猫“平平”“明明”飞赴爱尔兰共和国首都都柏林，进行为期100天的展览。16日，爱尔兰举行了中国大熊猫展出的开展仪式。6月21日，爱尔兰共和国总统帕特里克·希勒里以及爱尔兰财政部部长、林业部部长，苏联、美国、法国、土耳其等几十个国家的外交官员及其家属，也前来观看大熊猫。展出期间，大熊猫吸引了世界80多个国家和地区的观众，创造了门票神话。

On June 11, 1986, Ping-Ping and Ming-Ming the giant pandas flew to Dublin, the capital of the Republic of Ireland, for a 100-day exhibition. A ceremony for the launch of the Chinese giant panda exhibition was held in Ireland on June 16. On June 21, Dr. Patrick John Hillery, President of the Republic of Ireland, and the Irish Ministers of Finance, Energy and Forestry, etc., as well as the diplomats and their family members from the Soviet Union, the United States, France, Turkey and other countries, came to watch the pandas. During the exhibition, the giant pandas attracted a large number of audience from more than 80 countries and created a ticket myth.

大熊猫到达爱尔兰后，在当地引起了强烈反响。据说，爱尔兰就是否施行《离婚法》进行表决时，爱尔兰最具影响力的报刊《爱尔兰独立报》头版刊登了两只熊猫并排坐在一起的画面，并用大幅标题写道“100%的大熊猫是不离婚的”。这篇文章和照片在爱尔兰引起了强烈反响，最终有67.5%的人投票反对，没有通过《离婚法》。事后，有些爱尔兰人风趣地说：“今年的离婚法没有通过，是因为熊猫不离婚的缘故。”此次大熊猫出访爱尔兰，不仅加深了两国人民之间的友谊，也进一步扩大了中国在爱尔兰民众心中的良好形象。

After the Pandas arrived in lreland, they received a warm welcome there. It was said that when Ireland voted on whether to implement the Divorce Act. the Irish Independent, one of Ireland's most influential newspapers, ran a front-page picture of two pandas sitting side by side with the headline “100% of pandas are not divorced”. The article and photo reacted strongly in Ireland, with 67.5% voting against the divorce law. Later, some Irish people joked: “This year's divorce law was not passed because the pandas don’t divorce.” The giant pandas’ visit to Ireland brought the Irish people friendship of the Chinese people, expanded the influence of China's popularity in the Irish people, and enhanced the friendly relations between the two peoples.

“文文”“奔奔”助力申奥

Wen–Wen & Ben–Ben Helped Applying for Olympic

1998年11月25日，中国政府决定申办第29届奥运会。2001年7月，国际奥委会在俄罗斯莫斯科举行投票活动，以确定2008年奥运会的举办国。根据以往的惯例，所有申办城市都要在投票地举办“文化周”活动，介绍申办城市的优势，争取选票。经过慎重考虑，中国政府最后决定将“大熊猫赴莫斯科动物园展览”作为北京文化周的内容之一。

On November 25, 1998, the Chinese government decided to apply for the right to host the 29th Olympic Games. In July 2001, the IOC held a vote in Moscow, Russia, to determine the host country of the 2008 Olympic Games. According to the past practice, all the bidding cities should hold “Culture Week” activities in the voting place to introduce the advantages of the bidding cities and win the votes. After careful consideration, the Chinese government finally decided to include the “giant panda exhibition in the Moscow Zoo” as one of the contents of the Beijing Culture Week.

2001年6月13日，大熊猫“奔奔”和“文文”作为文化使者，带着中国人民的重托奔赴俄罗斯，为中国的申奥活动加油助威。大熊猫在莫斯科动物园一亮相，就引起了轰动，成为“北京文化周”的焦点，为北京申奥成功奠定了良好的基础。2001年7月13日，北京申奥成功，大熊猫们也完成了它们的历史使命。

On June 13, 2001, Ben-Ben and Wen-Wen the giant pandas, as cultural emissaries, went to Russia with the great trust of the Chinese people to cheer for China's Olympic bid. As soon as the giant pandas appeared at the Moscow Zoo, they caused a sensation and became the focus of the Beijing Culture Week, laying a good foundation for the success of Beijing's Olympic bid. On July 13, 2001, Beijing successfully bid for the Olympic Games, and the giant pandas also completed their historical mission.

第四篇
科研助手

根据1975年的《濒危野生动植物种国际贸易公约》，外国动物园只能以租借的方式，以科学研究的名义获得熊猫。于是，中国野生动物保护协会和中国动物园协会与国际动物保护机构经过两年磋商，达成合作研究协议。30年来，一批又一批大熊猫走出国门参与合作研究，它们把最好的年华留在了异国他乡。这样的合作，不仅有利于对大熊猫进行更为全面的研究，还进一步加强了中国与世界各国的交流与合作，增进了世界对中国的理解。

Chapter IV Research Assistant

According to the 1975 *Convention on International Trade in Endangered Species*, foreign zoos can only acquire pandas on loan from China, under the name of scientific research. Therefore, after two years of consultation, the China Wildlife Conservation Association (CWCA), Chinese Association of Zoological Gardens and the International Animal Protection Agency reached a joint research agreement. Over the past 30 years, batches of giant pandas went abroad to participate in the cooperative research and left their best time in foreign countries. Such cooperation is not only conducive to a more comprehensive research on the giant panda, but also further strengthens the exchanges and cooperation between China and other countries in the world, and enhances the world's understanding of China.

“永明”“蓉浜”“梅梅”爱撒日本

Yong-Ming, Rong-Bin & Mei-Mei Lived in Japan

1994年，中日签订协议，共同开展“大熊猫国际合作繁殖计划”，这也是首个大熊猫国际合作项目。1994年9月6日，大熊猫“永明”和“蓉浜”以“科研交流大使”的身份，开始了在日本和歌山白浜野生动物园的旅居生活。

In 1994, China and Japan signed an agreement to jointly launch the International Cooperative Giant Panda Breeding Program, which is also the first international cooperation project on giant pandas. On September 6, 1994, Yong-Ming and Rong-Bin the pandas began to live in the Adventure World in Shirahama, Wakayama, Japan as Ambassador for Scientific Research Exchange.

2000年7月7日，已怀孕的大熊猫“梅梅”踏足日本并于9月6日生下“良浜”。这是中日在国外合作繁育成活的第一只大熊猫，掀开了异国成功繁育大熊猫的历史新篇章。截至2019年11月，大熊猫“永明”的孙子辈总数已超过20只，还有3个曾孙，以一己之力创下了大熊猫海外合作繁殖最多的纪录，在大熊猫国际合作领域书写了辉煌一页。2022年12月17日，“永明”被任命为“中日友好特使”。

On July 7, 2000, Mei-Mei the pregnant giant panda set foot on the land of Japan and gave birth on September 6 to Liang-Bin, which was the first giant panda bred cooperatively abroad by China and Japan, opening a new chapter in the history of successful breeding of pandas in foreign country. Yong-Ming has more than 20 grandchildren and 3 great-grandchildren, setting a record for the largest overseas breeding of giant pandas, and writing a brilliant page in the field of international cooperation on giant pandas. On December 17, 2022, Yong-Ming was appointed as the Special Envoy for China-Japan Friendship.

2023年2月22日，大熊猫“永明”和它的双胞胎女儿“樱浜”“桃浜”搭乘飞机返回中国。2月24日，为纪念近30年的大熊猫国际科研繁育合作，日本和歌山白浜野生动物园特意向成都大熊猫繁育研究基地赠送了以大熊猫“永明”“梅梅”和他们的孩子为原型的纪念碑。50多年来，大熊猫在中日关系发展方面做出了重大贡献，拉近了两国民众的心理距离，是中日民间交往的重要桥梁。

On February 22, 2023, giant panda Yong-Ming and his twin daughters returned to China by plane. On February 24, in order to commemorate the nearly 30 years of international cooperation, the Adventure World in Shirahama, Wakayama, Japan presented to Chengdu Research Base of Giant Panda Breeding a monument based on Yong-Ming, Mei-Mei and their child. Over the past 50 years, giant pandas have made significant contributions to the development of China-Japan relations, narrowing the psychological distance between the people of the two countries, and being an important bridge for the people-to-people exchanges between China and Japan.

成都大熊猫繁育研究基地“永明”“梅梅”和它们的孩子的雕塑
Sculpture of Yong-Ming, Mei-Mei & their child at Chengdu Research Base of Giant Panda Breeding
图片来源：熊猫基地

“莉莉”“明明”小栖韩国

Li-Li & Ming-Ming Sojourned in ROK

1992年8月24日，中国与韩国正式建立大使级外交关系。1994年为纪念中韩建交，中国同意向韩国出借大熊猫。9月20日，大熊猫“莉莉”和“明明”（即“川星”）赴韩国龙仁爱宝乐园，也就是现在“爱宝”和“乐宝”所住的地方。

On August 24, 1992, the People’s Republic of China and the Republic of Korea (ROK) officially established diplomatic relations at the ambassadorial level. In 1994, China agreed to lend giant pandas to ROK to commemorate the establishment of diplomatic ties between China and ROK. On September 20, Li-Li and Ming-Ming (Chuan-Xing) the pandas went to Everland in Yongin, where Ai-Bao and Le-Bao live later.

然而，大熊猫的此次韩国之旅未能如期完成。1997年，亚洲金融危机波及韩国，韩国经济受到巨大冲击。因无力继续承担租借和饲养大熊猫的费用，1999年年初，韩国不得不将大熊猫送回中国。但此次大熊猫的韩国之旅为后来中韩大熊猫合作研究奠定了坚实的基础。2009年7月19日，在中国饲养员的精心照料下，“莉莉”诞下一对龙凤胎。在接下来的几年里，“莉莉”又连续生下了好几个孩子。

However, the giant panda’s trip to ROK was not completed as scheduled. In 1997, the Asian financial crisis spread to ROK, and ROK economy was greatly impacted. Unable to afford the cost of renting and raising the pandas, ROK had to sent the pandas back to China in early 1999. But the trip of the pandas to ROK laid a solid foundation for the subsequent cooperative research on giant pandas between China and ROK. On July 19th, 2009, under the Chinese feeders’ considerate care, Li-Li gave birth to a pigeon pair. In the following years, Li-Li delivered several children in succession.

“嫣嫣”远赴德国
Yan-Yan Went to Germany

1994年，中华人民共和国国务院总理李鹏在访德期间宣布中国政府将租借一只大熊猫给德国。1995年，柏林市长艾伯哈德·迪普根访华并带回了租借来的雌性大熊猫“嫣嫣”，代替过世的“天天”。“嫣嫣”的出现，拯救了动物园的人气，“嫣嫣”被选为“柏林荣誉市民”，一时风光无限。

In 1994, Li Peng, Premier of the State Council of the People’s Republic of China, announced that the Chinese government would lease a giant panda to Germany. In 1995, the mayor of Berlin Eberhard Diepgen visited China and personally brought back the rented female giant panda Yan-Yan to take the place of Tian-Tian the giant panda. The appearance of Yan-Yan saved the popularity of the zoo. Yan-Yan was also elected as the Honorary Citizen of Berlin, which made her best-known.

“嫣嫣”到德国后，由于一些游客不合适的投喂，“嫣嫣”养成了不良的饮食习惯。按照合同，“嫣嫣”本应于2000年回国，但德国希望留下它，就与中国续签了2年合同，2002年到期时又续了一次，2007年到期。2007年3月26日，“嫣嫣”因肠梗阻去世。“嫣嫣”死亡后，德国动物园将其制成了标本。

After arriving in Germany, Yan-Yan developed bad eating habits due to inappropriate feeding by some tourists. According to the contract, Yan-Yan was supposed to return to China in 2000, but Germany wanted to keep it longer and renewed the contract with China for two years in 2000, and renewed the contract again in 2002 ,which expired in 2007. On March 26, 2007, Yan-Yan died of intestinal obstruction after drinking a hard liquor fed by tourists. After her death, the German zoo made her into a specimen.

“白云”“石石”“高高”开枝美国

Bai-Yun, Shi-Shi & Gao-Gao Propagated in American

1994年，中国野生动物保护协会与美国圣地亚哥动物园签订为期12年的合作研究协议。1996年9月10日，大熊猫“白云”和“石石”抵达圣地亚哥动物园。1999年8月21日，“白云”和“石石”生下一只大熊猫宝宝，取名“华美”，寓意着中美人民之间的深厚友谊。“华美”是第一只在美国出生并存活下来的大熊猫，还是首只在国外出生后回到中国的大熊猫，有着非同一般的象征意义。

In 1994, the China Wildlife Conservation Association signed a 12-year cooperative research agreement with San Diego Zoo. On September 10, 1996, Bai-Yun and Shi-Shi the giant pandas arrived at the San Diego Zoo. On August 21, 1999, Bai-Yun and Shi-Shi gave birth to a giant panda baby, named Hua-Mei, symbolizing the profound friendship between the Chinese and American people. Hua-Mei is the first giant panda born alive in the United States, and also the first giant panda returned to China, full of extraordinary symbolic meaning.

2003年1月15日，大熊猫“高高”接替“石石”赴美。2008年12月18日，中美双方签署了为期五年的延期合作协议，到期后，于2013年又续签了五年的合作协议。“白云”与“高高”陆续生下“美生”“苏琳”“珍珍”“云子”“小礼物”，成为名副其实的英雄母亲。目前，大熊猫“高高”“白云”及6只幼仔都已先后回到中国。

On January 15, 2003, the giant panda Gao-Gao went to the United States to take the place of Shi-Shi. On December 18, 2008, China and the United States signed a five-year extension agreement. The two sides renewed a five-year cooperation agreement in 2013. Bai-Yun and Gao-Gao have given birth to Mei-Sheng, Su-Lin, Zhen-Zhen, Yun-zi and Xiao-Liwu successively, becoming a veritable heroic mother. At present, all giant pandas, including Gao-Gao, Bai-Yun and six cubs, have returned to China.

“伦伦”“洋洋”散叶美国

Lun–Lun & Yang–Yang Procreated in American

1999年，成都大熊猫繁育研究基地与美国亚特兰大动物园及美国佐治亚理工学院开展“国际合作繁殖计划”。根据协议，大熊猫“九九”和“华华”将旅居亚特兰大动物园10年。“九九”因生于9月9日而得名，后被荷兰一环保组织认养，改名“洋洋”，寓意“广阔的小海洋”。“华华”后改名“伦伦”，它比洋洋大半个月，非常活跃，且反应敏捷。

In 1999, Chengdu Research Base of Giant Panda Breeding and Zoo Atlanta and Georgia Institute of Tech nology launched the International Cooperative Breeding Program. According to the agreement, Jiu-Jiu and Hua-Hua the giant pandas would stay at the Zoo Atlanta for 10 years. Jiu-Jiu was born on September 9 and got the name which means double Nine. It was adopted by a Dutch environmental organization and changed its name to Yang-Yang, meaning a vast small ocean. Hua-Hua, which was renamed Lun-Lun, is older than Yang-Yang for half a month, active and responsive.

1999年11月5日，经过17小时旅行，大熊猫“洋洋”和“伦伦”飞抵亚特兰大市。上午9时30分，当画有两只大熊猫图案的专机滑向停机坪时，两台喷水车喷出数十米高的水柱，并形成一个水门，亚特兰大以这种独特的方式迎接大熊猫来美定居。中国驻休斯敦总领事吴祖荣、亚特兰大市长比尔·坎贝尔和亚特兰大动物园园长特里·梅普尔等200多人，在亚特兰大哈茨菲尔德-杰克逊国际机场热烈欢迎中国大熊猫的到来。11月20日，大熊猫正式与公众见面，成为亚特兰大的“明星夫妇”。

On November 5, 1999, after a 17-hour trip, Yang-Yang and Lun-Lun the giant pandas arrived in Atlanta. At 9:30 a.m., when the plane with two pandas slid toward the tarmac, two water sprinklers ejected dozens of meters high and formed an arched watergate. In this unique way, Atlanta welcomes the giant pandas to settle down in the United States. More than 200 people, including Chinese Consul General Wu Zurong in Houston, Mayor of Atlanta Bill Campbell and Director of the Zoo Atlanta, warmly welcomed the arrival of Chinese pandas at the Hartsfield International Airport. On November 20, the pandas officially met with the public and became star couple in Atlanta.

2006年9月7日，“伦伦”顺利产下一只雄性幼崽“美兰”。2008年8月，“伦伦”再次产下一只雄性小熊猫“喜兰”。2010年11月3日，“伦伦”产下阿宝（又名“宝兰”）。2013年7月15日，伦伦产下双胞胎“美轮”“美奂”。2016年9月3日，伦伦再次产下双胞胎“雅伦”“喜伦”。在多次延期后，亚特兰大动物园于2024年5月17日宣布，“伦伦”“洋洋”及“雅伦”“喜伦”将于2024年年底返回中国。

On September 7, 2006, Lun-Lun successfully gave birth to a male cub Meilan. In August 2008, Lun-Lun gave birth to a male baby Xi-Lan. On November 3,2010, Lun-Lun gave birth to Po (also known as Bao-Lan). On July 15,2013, Lun-Lun gave birth to her twins Mei-Lun and Mei-Huan. On September 3,2016, Lun-Lun gave birth to twins Ya-Lun and Xi-Lun. After several delays, Atlanta Zoo announced on May 17,2024 that Lun-Lun, Yang-Yang and their kids Ya-Lun and Xi-Lun would go back to China by the end of 2024.

“爽爽”“锦竹”“龙龙”祈福日本

Shuang-Shuang, Jin-Zhu & Long-Long Prayed for Japan

1995年1月17日，日本关西地区发生7.3级地震，造成重大人员伤亡和财产损失。为抚慰灾区人民的心灵，提振恢复重建的信心，位于日本关西的神户市王子动物园向中国提出租借大熊猫的请求。1999年，中国野生动物保护协会与日本神户王子动物园签署为期10年的大熊猫保护研究合作协议。2000年7月16日，大熊猫“爽爽”和“锦竹”抵达神户王子动物园。

On January 17, 1995, a 7.3-magnitude earthquake occurred in Kansai, Japan, causing heavy casualties and property losses. The Kobe Oji Zoo in Kansai, Japan, made a request for the loan of giant pandas to soothe the hearts of the people in the disaster areas and boost the confidence in post-disaster recovery and reconstruction. In 1999, the China Wildlife Conservation Association signed a 10-year cooperation agreement on giant panda conservation and research with Kobe Oji Zoo. On July 16, 2000, Shuang-Shuang and Jin-Zhu the giant pandas arrived at the Kobe Oji Zoo.

“爽爽”和“锦竹”是在为地震灾区复兴祈愿背景下赴日的，因此，“爽爽”和“锦竹”抵达日本后，根据公众意愿分别改名为“兴兴”和“旦旦”，寓意为“兴旺安康，旦夕平遂”。由于某些原因，“兴兴”（“锦竹”）于2002年被送回中国，另一只雄性大熊猫“龙龙”接替“锦竹”来到神户，成为第二代“兴兴”。2010年6月9日，中日签署补充协议，延长大熊猫合作研究5年。2015年，中国同意再次延期5年。

Shuang-Shuang and Jin-Zhu went to Japan under the background of praying for the revival of the earthquake-stricken areas. Therefore, Shuang-Shuang and Jin-Zhu were renamed Xing-Xing and Dan-Dan according to the public will, meaning prosperous and health, safe and smooth all the time. In 2002, due to certain reason, Xing-Xing (Jin-Zhu) was sent back to China, and another male giant panda Long-Long came to Kobe, becoming the second Xing-Xing. On June 9, 2010, China and Japan signed a supplementary agreement to extend the giant panda cooperation research for five years. In 2015, China agreed to extend another five years.

2010年9月9日，“兴兴”（“龙龙”）死亡，神户王子动物园就只剩下“旦旦”。按照协议，“旦旦”将于2020年7月回国。但因故无法敲定运送熊猫的航班。2021年3月，“旦旦”被查出患有心脏病，需要进行药物治疗。2024年3月31日，“旦旦”病情恶化，经抢救无效死亡。

On September 9, 2010, the giant panda Xing-Xing (Long-Long) suddenly died, leaving only Dan-Dan in Kobe Oji Zoo. According to the agreement, Dan-Dan would return to China in July 2020. But due to unforeseen circumstances, it had been impossible to locate flights carrying the pandas. In March 2021, Dan-Dan was diagnosed with heart disease and needed drug treatment. On March 31, 2024, Dan-Dan's condition deteriorated and it died after rescue.

“美香”“添添”添丁美国

Mei-Xiang & Tian-Tian Multiplied in American

2000年中国野生动物保护协会与美国华盛顿国家动物园签署了为期10年的大熊猫繁殖研究合作协议。2000年年底，大熊猫“美香”和“添添”抵达华盛顿国家动物园。按照美国哥伦比亚广播公司的说法，“华盛顿的熊猫时间到了”，它们“开始了保卫全球的职责”。2011年协议到期后，双方又签署了为期5年的延期协议。

In 2000, CWCA signed a 10-year cooperation agreement on giant panda breeding with Smithsonian's National Zoo. At the end of 2000, Mei-Xiang and Tian-Tian arrived at the Smithsonian's National Zoo in Washington. According to CBS News, “It's Panda Time In Washington.” they “got right down to their globe-defending duties.” In 2011, the two sides signed a five-year extension agreement after the agreement expires.

2005年，“美香”和“添添”产下“泰山”。作为第一只在华盛顿国家动物园诞生并成长的大熊猫，“泰山”在美国可谓是“超级巨星”，被视为“华盛顿最重要的居民”。按照最初的约定，“泰山”2周岁时就应被送回中国，但在美方多次请求下，中国同意延长“泰山”留美时间。2010年“泰山”终于回到中国保护大熊猫研究中心碧峰峡基地。

In 2005, Mei-Xiang and Tian-Tian gave birth to Tai-Shan. As the first giant panda to be born and raised at the National Zoo in Washington, Tai-Shan is a superstar in the United States and is regarded as the most important resident of Washington. According to the original agreement, Tai-Shan should be sent back to China at the age of 2. However, at the repeated requests of the US, China agreed to extend its stay in the US. In 2010, Tai-Shan finally returned to the Bifengxia Base of the China Conservation and Research Center for Giant Panda.

“美香”和“添添”先后生下“宝宝”和“贝贝”。2020年8月21日，在中美双方科研人员的共同努力下，刚过完22岁生日的“美香”在华盛顿产下“小奇迹”。2020年12月7日，美国华盛顿国家动物园宣布，该园和中国野生动物保护协会再次续签合作协议，大熊猫“美香”和“添添”以及幼崽“小奇迹”将在美国生活到2023年年底。2023年11月9日，它们终于回到四川成都。

Mei-Xiang and Tian-Tian then gave birth to Bao-Bao and Bei-Bei successively. On August 21, 2020, with the joint efforts of Chinese and American researchers, Mei-Xiang, who had just celebrated her 22nd birthday, gave birth to Xiao-Qiji (Little Miracle) in Washington. The Smithsonian's National Zoo announced on December 7 that the zoo and the Chinese Wildlife Conservation Association have renewed their cooperation agreement, giant pandas Mei-Xiang, Tian-Tian and their newly born cub Little Miracle, would live in the U.S.A. until the end of 2023. And they finally arrived in Chengdu, Sichuan province on November 9, 2023.

"阳阳""龙徽""园园"增福奥地利
Yang-Yang, Long-Wei & Yuan-Yuan Bred in Austria

2002年，在经过近10年的协商后，中国野生动物保护协会终于与奥地利美泉宫动物园签署为期10年的大熊猫繁殖研究合作协议。2003年3月3日，大熊猫"阳阳"和"龙徽"前往世界上最古老的动物园奥地利维也纳的美泉宫动物园。在这间有着250年历史的动物园，"阳阳"与"龙徽"是最耀眼的明星。

In 2002, after nearly 10 years of negotiations, the Chinese Wildlife Conservation Association signed a 10-year cooperation agreement on giant panda breeding research with Tiergarten Schönbrunn finally. On March 3, 2003, Yang-Yang and Long-Hui the giant pandas went to Tiergarten Schönbrunn, an oldest zoo with 250 years in the world, becoming the brightest stars in the zoo.

旅奥期间，“阳阳”与“龙徽”分别于2007年、2010年和2013年产下三只雄性幼仔“福龙”“福虎”和“福豹”，并于2016年产下一对龙凤胎“福伴”和“福凤”，创造了欧洲圈养大熊猫自然交配生产的纪录。目前，5只大熊猫宝宝都已回到中国。2016年12月9日，16岁“龙徽”因肿瘤手术麻醉期间停止心跳在美泉宫动物园离世。曾给无数游客带来欢乐的大熊猫“龙徽”被制作成标本后送回中国，做出了最后的贡献。

During their stay, Yang-Yang and Long-Hui gave birth to three male cubs Fu-Long, Fu-Hu and Fu-Bao in 2007, 2010 and 2013 respectively, and a pigeon pair Fu-Ban and Fu-Feng, in 2016, setting a record for the natural mating production of captive giant pandas in Europe. And now the five pandas have been back to China. On December 9, 2016, the 16-year-old giant panda Long-Hui died at the Tiergarten Schönbrunn during anesthesia of tumor surgery. Long-Hui the giant panda, which has brought joy to countless tourists, was made into specimens and sent back to China, making its final contribution.

“阳阳”与“龙徽”感情深厚。“龙徽”去世后，“阳阳”深感孤独。为慰藉“阳阳”，美泉宫动物园向中国提出申请再租借一只大熊猫。2019年4月15日，大熊猫“园园”飞赴奥地利陪同“阳阳”，并于5月20日与公众见面。

Yang-Yang has deep feeling to Long-Hui. Without Long-Hui, Yang-Yang felt lonely. To comfort Yang-Yang, Tiergarten Schönbrunn has applied to China to rent another giant panda. On April 15, 2019, Yuan-Yuan the giant panda flew to Austria to accompany Yang-Yang and met with the public on May 20.

“林惠”“创创”结缘泰国

Lin–Hui & Chuang–Chuang Form Ties with Thailand

2003年，中国野生动物保护协会与泰国动物园管理协会签订为期10年的大熊猫繁殖研究合作协议。2003年10月12日，大熊猫“林惠”和“创创”抵达泰国清迈动物园，帮助开展合作研究，成为中泰两国的友好大使。

In 2003, CWCA and The Zoological Park Organization of Thailand signed a 10-year cooperation agreement on giant panda breeding research. On October 12, 2003, Lin-Hui and Chuang-Chuang the giant pandas arrived at the Chiang Mai Zoo in Thailand to help carry out the collaborative research and become the friendship ambassador between Thailand and China.

2009年5月27日，大熊猫“林惠”产下一只雌性大熊猫幼仔“林冰”，“林冰”于2013年9月28日返回中国。2015年6月，中泰双方签署大熊猫合作延期协议，将“创创”和“林惠”在泰国的居住期限延长至2023年年底。2019年9月16日，“创创”因慢性心力衰竭急性发作导致全身器官缺氧而突然在清迈动物园死亡，年仅19岁。2023年4月19日，旅居泰国清迈动物园的22岁雌性大熊猫“林惠”因病抢救无效离世。

On May 27, 2009, the giant panda Lin-Hui gave birth to a female cub Lin-Bing, who returned to China on September 28, 2013. In June 2015, China and Thailand signed an agreement to extend the cooperation, extending the stay of Chuang-Chuang and Lin-Hui in Thailand to the end of 2023. On September 16, 2019, Chuang-Chuang died suddenly at the age of 19 due to an acute attack of chronic heart failure. On April 19, 2023, Lin-Hui, a 22-year-old female giant panda living at the Chiang Mai Zoo, Thailand, died of illness.

“丫丫”“乐乐”旅居美国

Ya-Ya & Le-Le Sojourned in American

1999年4月，为开展大熊猫保护合作研究，在时任美国驻华大使尚慕杰的倾力帮助下，美国孟菲斯动物园成功与中国动物园协会达成从中国租借两只大熊猫的协议。2003年4月7日，大熊猫“乐乐”和“丫丫”抵达美国田纳西州孟菲斯动物园，开始10年的旅居生活。“丫丫”和“乐乐”的到来，使孟菲斯动物园成为全美第四家拥有大熊猫的动物园。2013年协议到期后，又延长了10年。

In April 1999, to conduct a cooperative research on giant panda conservation, with the help of James Sasser, then American Ambassador to China, the Memphis Zoo successfully reached an agreement with the Chinese Association of Zoological Gardens to lease two giant pandas from China. On April 7, 2003, Le-Le and Ya-Ya the giant pandas arrived at the Memphis Zoo in Tennessee for a 10-year living life. The arrival of Ya-Ya and Le-Le also made the Memphis Zoo the fourth zoo in the United States to have giant pandas. A further 10-year extension was signed after the 2013 agreement expired.

为了迎接“丫丫”和“乐乐”，孟菲斯动物园在2003年耗资1600万美元建成了极具中国传统文化特色、功能齐全的大熊猫场馆。“丫丫”和“乐乐”在美国当地受到众多民众的喜爱和追捧。它们到达美国的第一天，即被孟菲斯市长称为“孟菲斯的高光时刻”。为了庆祝大熊猫的到来，“丫丫”在2003年5月还登上了《时代》杂志儿童版的封面。

In order to welcome Ya-Ya and Le-Le, the Memphis Zoo built a fully functional giant panda venue full of Chinese traditional culture with a cost of $16 million in 2003. Ya-Ya and Le-Le are loved and sought after by many local people in the United States. The first day they arrived in the United States were called “the highlight moment of Memphis” by the mayor of Memphis. To celebrate the arrival of the giant pandas, Ya-Ya also appeared on the cover of *Times for kids* on May 2, 2003.

2003年5月2日“丫丫”登上《时代》杂志儿童版封面
Ya-Ya appears on the cover of *Time for Kids* on May 2, 2003

当地时间2023年2月1日早晨，大熊猫“乐乐”死于心脏病，终年25岁。4月8日上午，数百人出席了“乐乐”的吊唁活动。4月27日下午，大熊猫“乐乐”的遗体随“丫丫”一道抵达上海浦东国际机场。在过去的20年里，“丫丫”和“乐乐”吸引了数百万游客，是大熊猫模范大使和中美合作开展大熊猫保护的光辉象征。

On the morning of February 1, 2023, the giant panda Le-Le died of heart disease at the age of 25. On the morning of April 8, hundreds of people attended the condolence ceremony. On the afternoon of April 27, Ya-Ya and the remains of the dead Le-Le arrived at the Shanghai Pudong International Airport. Over the past 20 years, Ya-Ya and Le-Le have delighted millions of visitors, serving as exemplary ambassadors for their species and remaining a shining symbol of conservation partnership between America and China.

“花嘴巴”“冰星”燃情西班牙
Hua-Zuiba & Bing-Xing Loved in Spain

2007年6月24日至29日，应中华人民共和国国家主席胡锦涛邀请，西班牙国王胡安·卡洛斯一世对中国进行国事访问。访华期间，中国与西班牙两国政府确定大熊猫保护研究国际合作项目启动。2007年9月8日，大熊猫“花嘴巴”和“冰星”抵达马德里，11天后，西班牙王后索菲亚专程来到马德里动物园“接见”它们。

At the invitation of Chinese President Hu Jintao, King Juan Carlos I of Spain paid a state visit to China from June 24 to 29, 2007. During his visit, the Chinese and Spanish governments launched the giant panda international conservation cooperation project. On September 8, Hua-Zuiba and Bing-Xing the giant pandas arrived in Madrid. Eleven days later, Queen Sophia of Spain went to Madrid Zoo to meet them.

2010年9月7日，它们的双胞胎儿子“德德”和“阿宝”诞生。当年11月，西班牙王后索菲亚来到马德里动物园大熊猫馆探望大熊猫，并为熊猫幼崽喂奶。这是中国与国外开展大熊猫长期国际繁育合作计划以来，在海外产下的首对人工授精大熊猫双胞胎。之后，“花嘴巴”和“冰星”分别于2013年8月30日和2016年8月30日生下“星宝”和“竹莉娜”。

On September 7, 2010, their twin sons, De-De and A-Bao, were born. In November of that year, Queen Sophia of Spain came to Madrid Zoo to visit the pandas and fed the cubs. This is the first pair of artificial insemination panda twins born overseas since China launched a long-term international breeding cooperation program. Later, Hua-Zuiba and Bing-Xing gave birth respectively to Xing-Bao on August 30, 2013 and Chu-Lina on August 30, 2016.

2018年2月23日，西班牙马德里动物园与中方续签中西大熊猫合作研究协议，马德里动物园将继续保留一对大熊猫，直至2023年。2021年9月6日，“冰星”和“花嘴巴”产下双胞胎“久久”和“友友”。大熊猫是中国的国宝，也是中国与西班牙友好关系和文化交流的象征。

In February 23，2018, Madrid Zoo renewed the agreement with China, and would keep a pair of giant pandas until 2023. On September 6, 2021, Bing-Xing and Hua-Zuiba gave birth to twins Jiu-Jiu and You-You. The giant panda is a national treasure of China and a symbol of the high level of friendly relations and cultural exchanges between China and Spain.

“网网”“福妮”首探澳洲
Wang-Wang & Fu-Ni Explored in Australia

根据2007年中澳签署的一份为期10年的协议，中国将向阿德莱德动物园出租一对大熊猫用于两国开展大熊猫繁殖合作研究。自2009年11月27日起，大熊猫“网网”和“福妮”开始定居澳大利亚阿德莱德动物园，成为南半球的第一对大熊猫。2019年，中澳签订协议，将大熊猫的租期延长至2024年。

Under a 10-year agreement signed between China and Australia in 2007, China would lease a pair of giant pandas to Adelaide Zoo for cooperative breeding research between the two countries. Since November 27, 2009, Wang-Wang and Fu-Ni have settled in Adelaide Zoo in Australia, becoming the first pair of pandas in the southern hemisphere. In 2019, China and Australia signed an agreement to extend the lease term of the pandas until 2024.

中国驻澳大利亚大使肖千在首次正式访问澳大利亚时也来到了阿德莱德动物园参观。动物园园长伊莱恩·本斯特德热烈欢迎肖千到访，双方进行了友好交流。大熊猫作为中澳民间交流大使，增进了两国人民的相互理解和友好情感，为中澳两国民心相通发挥了积极作用。

The Chinese Ambassador to Australia, Xiao Qian, also visited the Adelaide Zoo during his first official visit to Australia. Elaine Bensted, the director of the zoo, warmly welcomed Xiao Qian and the two sides had friendly exchanges. The giant panda, as the ambassador for people-to-people exchanges between China and Australia, has enhanced the mutual understanding and friendship between the two peoples, and play a positive role in the communication between China and Australia.

“比力”“仙女”栖身日本

Bi-Li & Xian-Nü Left for Japan

2008年4月30日凌晨，大熊猫“陵陵”的去世，震惊了日本。日本首相福田康夫第二天即向中国提出租借大熊猫的请求。2008年5月6日，中华人民共和国国家主席胡锦涛访日，当晚，在与福田康夫首相共进晚餐时，胡锦涛表示对大熊猫“陵陵”的不幸去世深感惋惜，理解日本广大民众对大熊猫的喜爱，也注意到福田首相对此事的高度重视，中国同意向日本提供大熊猫一对，以供研究之用。

The death of giant panda Ling-Ling in the early hours of April 30, 2008 shocked Japan. Japanese Prime Minister Yasuo Fukuda made a request to China for a panda loan the next day. On May 6, 2008, Chinese President Hu Jintao visited Japan. At dinner that day with Prime Minister Yasuo Fukuda, Hu Jintao said he was deeply sorry for the death of the giant panda Ling-Ling, understood Japanese public's love for giant pandas, and noted the great importance prime minister Fukuda attached to the matter. China agreed to provide Japan with a pair of giant pandas for research.

2010年5月，由上野动物园大熊猫饲养员、兽医等6人组成的“大熊猫相亲团”前往中国四川省和广东省的两处大熊猫基地挑选大熊猫。身强力壮的“比力”和憨态可掬的“仙女”瞬间得到“相亲团”的一致认可。2011年2月22日，大熊猫“比力”和“仙女”抵达东京上野动物园。2011年2月，抵达上野动物园后，“比力”因为很有活力改名“力力”，“仙女”因为纯洁天真改名“真真”。它们先后产下“香香”及双胞胎“晓晓”和“蕾蕾”。“香香”是上野动物园独一无二的“大明星”。2023年2月21日，大熊猫“香香”返回中国。

In May 2010, a group of six, including panda keepers and veterinarians, went to two panda bases in Sichuan and Guangdong provinces to choose pandas. Bi-Li, a strong male panda, and Xian-Nü, a naive and charming female, attracted the group instantly. Bi-Li and Xian-Nü the giant pandas arrived at the Ueno Zoo in Tokyo on February 22, 2011. Bi-Li was renamed to Li-Li to emphasize his playful vitality. Xian-Nü was changed to Shin-Shin, referring to purity and innocence. They gave birth to Xiang-Xiang and the twins, Xiao-Xiao and Lei-Lei. Xiang-Xiang is the unique superstar in Ueno Zoo. On February 21, 2023, Xiang-Xiang returned to China.

"甜甜""阳光"联结中英
Tian–Tian & Yang–Guang United China and UK

2011年1月10日，在中华人民共和国国务院副总理李克强和英国副首相尼克·克莱格的共同见证下，中国野生动物保护协会与苏格兰皇家动物学会在伦敦正式签署了为期10年的大熊猫保护研究合作协议。2011年12月4日，大熊猫"甜甜"和"阳光"抵达苏格兰爱丁堡皇家动物园，开始了为期10年的"留英"生涯。

On January 10, 2011, in the presence of Chinese Vice Premier Li Keqiang and British Deputy Prime Minister Nick Clegg, China Wildlife Conservation Association and the Royal Zoological Society of Scotland officially signed a 10-year cooperation agreement on giant panda conservation and research in London. Tian-Tian (Sweet) & Yang-Guang (Sunshine) the giant pandas, symbolizing the friendship between China and Britain, arrived at the Royal Zoo, Edinburgh, Scotland, on December 4, 2011 for a 10-year stay in the UK.

在大熊猫“甜甜”和“阳光”落户英国之后，爱丁堡动物园就成为全球第10家拥有大熊猫的中国境外动物园。作为英国仅有的两只大熊猫，它们极受游客欢迎。按照最初的协议，它们本应于2021年返回中国，但因故协议延长了两年。2023年12月4日，“甜甜”和“阳光”启程返回中国并于12月5日抵达成都。至此，英国境内再无大熊猫。

The Zoo in Edinburgh became the 10th zoo where people could meet pandas outside of China after Tian-Tian and Yang-Guang the pandas settled in the UK. As the only two giant pandas in the UK, they are extremely popular with tourists. Under the first agreement, they were due to return to China in 2021, but the agreement was extended for two years due to unforeseen circumstances. On December 4, 2023, Tian-Tian and Yang-Guang left for China and arrived in Chengdu on December 5. So far, there are no giant pandas in the UK.

“欢欢”“圆仔”安身法国

Huan–Huan & Yuan–Zai Stayed in France

自2005年以来，法国博瓦勒野生动物园多次到中国成都大熊猫繁育研究基地就双方合作进行商谈。2011年12月3日，中法两国在法国驻华大使馆正式签署为期10年的大熊猫保护与研究合作协议。2012年1月15日，在中法建交48周年之际，大熊猫“圆仔”和“欢欢”飞赴法国博瓦勒野生动物园。自2000年1月20日大熊猫“燕燕”去世以来，这是大熊猫第一次踏足法国！

Since 2005, Beauva Zoo in France has visited Chengdu Research Base of Giant Panda Breeding in China for many times to discuss bilateral cooperation. On December 3, 2011, China and France officially signed a 10-year cooperation agreement on giant panda protection and research. On January 15, 2012, on the occasion of the 48th anniversary of the establishment of diplomatic relations between China and France, Yuan-Zai and Huan-Huan the giant pandas flew to Zoo Parc de Beauval in France. This is the first time a giant panda has set paw on French soil since the giant panda Yan-Yan died on January 20, 2000!

2017年8月4日，“欢欢”产下一只雄性大熊猫“圆梦”。圆梦是首只在法国出生的大熊猫，它似乎比它的父母更受欢迎。法国“第一夫人”布丽吉特·马克龙亲自为它取名，总统马克龙甚至选择在动物园和“圆梦”一起度过自己40岁的生日。2021年8月2日，“欢欢”又产下一对双胞胎幼崽“欢黎黎”“圆嘟嘟”。

On August 4, 2017, Huan-Huan gave birth to a male giant panda Yuan-Meng who became the first giant panda born in France. Yuan-Meng seems to be more popular than it’s parents. The French first Lady Brigitte Macron named it herself, and French President Emmanuel Macron even chose to spend his 40th birthday with Yuan-Meng at the zoo. On August 2, 2021, Huan-Huan also gave birth to a twin cubs Huan-Lili and Yuan-Dudu.

应中华人民共和国国家主席习近平邀请，法国总统马克龙于2023年4月5日至7日对中国进行国事访问，博瓦勒野生动物园园长鲁道夫·德洛尔随马克龙一道访华。2023年4月11日，法国宣布续租大熊猫成功，“欢欢”“圆仔”将留法至2027年。2023年7月26日大熊猫“圆梦”被送回中国。临行前，法国总统夫人布丽吉特专程到动物园为“圆梦”送行。

At the invitation of Chinese President Xi Jinping, French President Emmanuel Macron paid a state visit to China from April 5 to 7, 2023. Rudolf Delor, director of Zoo Parc de Beauval, accompanied. On April 11, 2023, France announced that the giant pandas were successfully renewed, and Huan-Huan and Yuan-Zai would stay in France until 2027. On July 26, 2023, Yuan-Meng returned to China. Brigitte, the wife of the French president, came to the zoo to say goodbye to Yuan-Meng.

“武杰”“沪宝”暂住新加坡
Wu-Jie & Hu-Bao Settled in Singapore

2009年11月，为庆祝中国和新加坡建交20年，中新双方共同签署了大熊猫保护研究合作协议。经过近3年的准备，2012年9月6日，大熊猫“武杰”和“沪宝”来到新加坡河川生态园，受到了当地民众的喜爱。经过6个月征集名字，“武杰”和“沪宝”分别改名为“凯凯”和“嘉嘉”。因为“凯”字含有“胜利”的意思，象征着中新关系20年的丰硕成果；“嘉”字意味着美好，反映了良好的中新关系。

In November 2009, to celebrate the 20-year establishment of diplomatic ties between China and Singapore, China and Singapore signed an agreement on panda cooperation and research. After nearly three years of preparation, Wu-Jie and Hu-Bao the giant pandas came to Singapore River Safari on September 6, 2012, and were loved by the local people. After six months of name soliciting, Wu-Jie and Hu-Bao were renamed Kai-Kai and Jia-Jia respectively. Because the Chinese character Kai contains the meaning of victory, symbolizing the fruitful achievements of 20 years; and the Chinese character Jia means beautiful, reflecting the good relations between China and Singapore.

2021年8月14日，在经过7年的尝试后，“凯凯”和“嘉嘉”终于有了自己的第一个宝宝“叻叻”。“叻叻”成为新加坡最受关注的动物宝宝之一，受到新加坡人民的喜爱。新加坡总理李显龙还通过社交媒体发文祝贺。“叻叻”也进一步加深了中新两国人民的感情。

On August 14, 2021, after seven attempts, Kai-Kai and Jia-Jia finally had their first baby Le-Le, who has become one of the most widely discussed animal babies in Singapore, and is loved by the people of Singapore. Singaporean Prime Minister Lee Hsien Loong also congratulated via social media. Le-Le further deepens the feelings of the Chinese and Singapore peoples.

“叻叻”与自己的“登机牌”合影
Le-Le poses with his “Boarding Pass”

2023年12月13日，新加坡河川生态园举行了一场特别的欢送会，主角是当地的动物明星——首只在新加坡出生的大熊猫“叻叻”。数千名当地居民来到这里，在“叻叻”回中国前见它一面。“叻叻”于2024年1月16日乘飞机回到中国，入住华蓥山大熊猫野化放归培训基地。

On December 13, 2023, the Singapore River Safari held a special farewell party featuring the local animal star—the first Singapore-born giant panda Le-Le. Thousands of local residents went to see him before he returned to China. Le-Le returned to China by plane on January 16, 2024 and move to the Huaying Mountain Giant Panda Wild Training Base.

“大毛”“二顺”侨居加拿大

Da-Mao & Er-Shun Lived in Canada

2012年2月11日，加拿大总理哈珀访问中国，中加签署为期10年的大熊猫保护与研究合作协议。2013年3月25日，大熊猫“大毛”和“二顺”飞赴加拿大多伦多动物园。加拿大总理哈珀携夫人劳琳及其他政要在多伦多皮尔逊国际机场迎接来自中国的大熊猫。哈珀总理在机场发表了热情洋溢的讲话。他说，大熊猫的到来有助于中加两国人民相互加深了解。

On February 11, 2012, Canadian Prime Minister Stephen Harper visited China. China and Canada signed a 10-year cooperation agreement on giant panda conservation and research. On March 25, 2013, Da-Mao and Er-Shun the giant pandas flew to Toronto Zoo in Canada. Canadian Prime Minister Stephen Harper, his wife Lauine and his dignitaries welcomed the arrival of Chinese giant pandas at the Toronto Pearson International Airport. Harper delivered a glowing speech at the airport. He said that the arrival of the giant pandas would help the Chinese and Canadian people to deepen their mutual understanding.

它们先后在多伦多动物园和卡尔加里动物园各生活 5 年。2015年10月13日，“二顺”在加拿大多伦多动物园产下一对龙凤胎熊猫宝宝，后被命名为“加盼盼”和“加悦悦”。2016年3月7日，加拿大总理贾斯廷·特鲁多出席了命名仪式。

The two pandas lived at the Toronto Zoo and Calgary Zoo for five years successively. On October 13, 2015, Er-Shun gave birth to a boy-and-girl twins at the Toronto Zoo. Later they were named Jia-Panpan and Jia-Yueyue. Then Canadian Prime Minister Justin Trudeau also attended the naming ceremony on March 7, 2016.

2018年3月，4只大熊猫一道迁居卡尔加里动物园。为确保大熊猫吃上美味的竹子，卡尔加里动物园一直坚持从中国进口新鲜竹子。但有一段时间竹子供应困难。为此，加拿大不得不将大熊猫提前返还中国。2020年11月29日，“大毛”和“二顺”抵达成都，结束了在加拿大的旅居生活。“加盼盼”和“加悦悦”已于2020年1月12日先行返回中国。

In March 2018, the four giant pandas moved to the Calgary Zoo together. Calgary Zoo had been importing fresh bamboo from China to ensure the pandas eat delicious bamboo. However, for a while, there has been difficulty in obtaining bamboo supplies. Canada had to return the pandas to China in advance. On November 29, 2020, Da-Mao and Er-Shun arrived in Chengdu, ending their sojourn life in Canada. Jia-Panpan and Jia-Yueyue returned to China first on January 12, 2020.

"福娃""凤仪"南下马来

Fu-Wa & Feng-Yi Head for Malaysia

2012年，马来西亚总理纳吉布访问中国南宁，中国同意与马来西亚开展大熊猫保护合作研究。2013年6月15日，中国野生动物保护协会与马来西亚野生动物和国家公园局签署大熊猫保护合作研究协议，以庆祝中马建交40周年。2014年5月21日大熊猫"福娃"和"凤仪"抵达马来西亚国家动物园，开启了为期10年的旅居生涯。

In 2012, Malaysian Prime Minister Najib Razak visited Nanning, China, and China agreed to carry out cooperative research on giant panda conversation with Malaysia. On June 15th, 2013, CWCA signed an agreement on cooperative research on giant panda conversation with the Wildlife and the National Park Bureau of Ministry of Natural Resources and Environment (Malaysia), celebrating the 40th anniversary of the establishment of diplomatic relations between China and Malaysia. On May 21, 2014, Fu-Wa and Feng-Yi the giant pandas arrived at the Malaysia's National Zoo for their 10-year sojourn life.

“福娃”和“凤仪”抵达马来西亚后，受到了马来西亚自然资源与环境部长帕拉尼维尔和中国驻马来西亚大使黄惠康的热烈欢迎。马来西亚总理纳吉布在出席熊猫馆开馆仪式时宣布将“福娃”和“凤仪”改名为“兴兴”和“靓靓”。目前，这对明星夫妻已连生三个女儿。2017年11月5日，出生于2015年的长女“暖暖”返回中国。2023年8月29日晚，大熊猫“谊谊”和“升谊”乘飞机返回中国并于8月30日平安抵达成都。

Upon arrival, Fu-Wa and Feng-Yi were warmly welcomed by Malaysian Natural Resources and Environment Minister G. Palanivel and Chinese Ambassador to Malaysia Huang Huikang. At the opening ceremony of the panda house, Malaysian Prime Minister Najib Razak renamed Fu-Wa and Feng-Yi as Xing-Xing and Liang-Liang. So far, this couple have had three daughters. On November 5, 2017, the eldest daughter Nuan-Nuan, born in 2015, returned to China. On the evening of August 29, 2023, the second daughter Yi-Yi, born in 2018, and the youngest daughter Sheng-Yi, born in 2021, flew back to China and arrived in Chengdu safely on August 30.

2023年8月23日，马来西亚举办“2023大熊猫生日会”。会上，马来西亚自然资源、环境与气候变化部副秘书长阿卜杜勒·瓦希德致辞：“大熊猫作为中国特使，促进了两国之间的理解和密切合作……体现了马中之间的密切友谊和合作。”马来西亚国家动物园副主席罗斯利表示：“这些大熊猫不仅是可爱的标志，更是国际友好的使者，象征着两国之间经久不衰的友谊。它们俘获了人们的心灵，跨越了文化的鸿沟，激励人们为了更美好的未来而共同努力。”

Malaysia held the Giant Panda Birthday Party 2023 on August 23, 2023. At the party, Abdul Wahid Abu Salim, deputy secretary-general of the Natural Resources, Environment and Climate Change Ministry (NRECC), said: “The giant pandas serve as a special envoy from China, fostering understanding and close cooperation between our nations...This project has come to embody the close friendship and collaboration between Malaysia and China...” Zoo Negara Deputy President Rosly Rahmat Ahmat Lana said: “These pandas are not just an adorable icon, they are ambassadors of goodwill and symbols of the enduring friendship between our countries. They have captured our hearts and minds, bridging cultural divides and inspiring us to work together for a better future.”

“好好”“星徽”投宿比利时

Hao-Hao & Xing-Hui Put up in Belgian

2013年9月，中国野生动物保护协会与比利时埃诺省布吕热莱特市天堂动物园签署为期15年的大熊猫合作研究协议。2014年2月23日，大熊猫“星徽”和“好好”抵达天堂动物园。比利时首相迪吕波前往布鲁塞尔国际机场迎接。

In September 2013, CWCA and Pairi Daiza in Brugelette, Province de Hainaut, Belgium officially signed a 15-year cooperative research agreement on giant pandas. On February 23, 2014, Xing-Hui and Hao-Hao the giant pandas arrived in Pairi Daiza. Belgian Prime Minister Elio Di Rupo received the pandas at the Brussels Airport.

2014年3月30日下午，中华人民共和国国家主席习近平和比利时国王菲利普共同出席比利时天堂公园大熊猫园开园仪式。习近平和夫人彭丽媛抵达时，受到菲利普国王和玛蒂尔德王后、比利时首相迪吕波热情迎接。

On the afternoon of March 30, 2014, Chinese President Xi Jinping attended the opening ceremony of the giant panda garden in Pairi Daiza with King Philip. When Xi Jinping and his wife Peng Liyuan arrived, they were greeted and warmly welcomed by King Philip, Queen Mathilde and Belgian Prime Minister Elio Di Rupo.

天堂动物园里的中国园
The China Garden in the Pairi Daiza
图片来源：Sollyn 供图

天堂动物园是欧洲最出名的动物园之一，被誉为“比利时最美动植物园”，其中大熊猫居住的中国园是目前整个欧洲最大的中国式园林，深受游客喜爱。2016年6月1日，“好好”生下了第一只熊猫宝宝“天宝”；2019年8月8日，“好好”又诞下大熊猫双胞胎“宝弟”和“宝妹”。

Pairi Daiza is the most famous zoo in Europe, known as “the most beautiful animal and botanical garden in Belgium”. The Chinese garden where giant pandas live is the largest Chinese style garden in Europe, which is deeply loved by tourists. Hao-Hao gave birth to her first baby panda, Tian-Bao, on June 1, 2016, and gave birth to her giant panda twins Bao-Di and Bao-Mei on August 8, 2019.

“华妮”“园欣”东去韩国

Hua–Ni & Yuan–Xin Left for ROK

2014年7月，中华人民共和国国家主席习近平对韩国进行国事访问，与韩国总统朴槿惠达成共识，安排一对大熊猫到韩国参加联合科学研究。2016年3月3日，大熊猫“华妮”和“园欣”抵达韩国并入住爱宝乐园，开始为期15年的“旅韩”生活。韩国由此成为第14个拥有大熊猫的国家。时隔22年之后，韩国民众终于有机会再睹中国大熊猫的风采了。

Chinese President Xi Jinping paid a state visit to the ROK in July 2014. During a summit meeting with President Park Geun-hye, Xi Jinping promised to lend a pair of giant pandas to ROK for joint research. On March 3, 2016, Hua-Ni and Yuan-Xin the pandas arrived in ROK and live in Everland, starting their 15-year life in ROK. This made ROK the 14th country to have giant pandas. It’s also the second time after 22 years that ROK have a chance to get a glimpse of the Chinese giant pandas.

大熊猫抵达韩国后，韩国为大熊猫广泛征集名字，最终确定给两个小伙伴改名为“爱宝”和“乐宝”，寓意大熊猫能够给中韩两国人民带来快乐与幸福，增进两国的友好关系。“爱宝”和“乐宝”在韩国得到了精心照料。2020年7月20日，“爱宝”平安诞下一只雌性幼仔“福宝”。

After the pandas arrived in Republic of korea (ROK), the goverment widely solicited names for them, and finally decided to renamed them Ai-Bao and Le-Bao, meaning that the pandas could bring happiness and pleasure to the Chinese and Korean peoples and enhance the friendly relations between the two countries. Ai-Bao and Le-Bao are well cared for in ROK. On July 20, 2020, Ai-Bao gave birth to a female cub Fu-Bao.

“武雯”“星雅”逐芳荷兰

Wu-Wen & Xing-Ya Chased in Dutch

早在2000年，荷兰政府就萌生了向中国租借大熊猫的想法，当时的首相还亲自给中国领导人写信请求此事，荷兰王室、政界、商界、学界也都曾为租借大熊猫而四处奔走。终于在2015年10月，荷兰国王威廉-亚历山大访华期间，中国方面终于同意借给荷兰一对大熊猫，用于大熊猫合作研究项目。

As early as the year 2000, the Dutch government came up with the idea of renting pandas from China. The prime minister personally wrote a letter to the Chinese leader, and the Dutch royal family, political, business and academic circles also ran around for the loan of giant pandas. Finally, during the visit of King Willem-Alexander of the Netherlands in October 2015, China agreeing to lend a pair of giant pandas to the Netherlands for joint research projects on giant panda.

为了让大熊猫有家的感觉，欧维汉兹动物园不惜耗费巨资700万欧元，打造了一座9000多平方米的中国古代皇家宫殿风格住所。住所的一砖一瓦都是从中国空运过去的，堪称“史上最豪华的熊猫馆”。

欧维汉兹动物园“中国熊猫馆”全景模型
Pandasia model in Ouwehands Dierenpark

To make the pandas feel at home, the Zoo spent 7 million euros to build a 9,000-square-meter Chinese traditional palace style residence “Pandasia”. Every brick and tile of the residence, the most luxurious panda house in history, was flown from China.

2017年4月12日，在中荷建交45周年之际，大熊猫“星雅”和“武雯”抵达郁金香之国荷兰，受到荷兰首相马克·吕特的迎接。两只大熊猫将在荷兰欧维汉兹动物园熊猫馆停留15年，而荷兰也成为第7个拥有大熊猫的欧洲国家。

On April 12, 2017, on the occasion of the 45th anniversary of the establishment of diplomatic ties between China and the Netherlands, Xing-Ya and Wu-Wen the giant pandas arrived in Netherlands, the land of tulips, and were welcomed by the Dutch Prime Minister Mark Rutte. The two pandas would stay in Ouwehands Dierenpark in the Netherlands for 15 years, making the Netherlands the seventh European country that hosts this endangered and adorable black and white bear.

“星雅”和“武雯”没有辜负荷兰人民的厚爱，于2020年5月1日生下女儿“梵星”，为荷兰人民带去了更多欢乐。“梵星”的名字分别取自荷兰画家凡·高及其名作《星空》，以及它的父亲“星雅”。2023年9月27日，欧维汉兹动物园为大熊猫“梵星”举办隆重的欢送仪式，数百名荷兰民众依依不舍，挥手告别。

Xing-Ya and Wu-Wen did not live up to the love of the Dutch people, and gave birth to their daughter Fan-Xing on May 1, 2020, bringing more joy to the Dutch people. The name Fan-Xing is taken from the Dutch painter Van Gogh and his famous work *The Starry Night*, and the father panda Xing-Ya. On September. 27, 2023, Ouwehands Dierenpark held a grand farewell ceremony for Fan-Xing, and hundreds of Dutch people came to wave goodbye.

"娇庆""梦梦"安居德国

Jiao-Qing & Meng-Meng Settled in Germany

自2012年8月22日大熊猫"宝宝"离世后，德国一直希望能再从中国租借大熊猫。2015年10月，德国总理默克尔在第8次访华时为德国带回了好消息：两年内德国将从中国租借一对熊猫。2017年4月28日，中国野生动物保护协会与德国柏林动物园签署了15年的大熊猫保护研究合作协议。在盼望了五年之后，德国公众终于又能够再次在家乡看到大熊猫这种神奇的动物了。

Since the giant panda Bao-Bao died on August 22, 2012, Germany had hoped to lease giant pandas from China again. German Chancellor Angela Merkel brought back good news for Germany during her eighth visit to China in October 2015 that Germany would lease a pair of pandas from China within two years. On April 28, 2017, China Wildlife Conservation Association and Zoo Berlin signed a 15-year agreement on giant panda conservation and joint research. After five years of expecting, the German public could finally see this magical animal, the giant panda, in their hometown again.

大熊猫“梦梦”和“娇庆”乘坐的专机抵达德国柏林机场
The plane carrying pandas "Meng Meng" and "Jiao-Qing" from China arrives at an airport of Berlin, capital of Germany, on June 24, 2017.
图片来源：新华社记者 单宇琦 摄

2017年6月24日，大熊猫“梦梦”和“娇庆”抵达德国。柏林市市长穆勒、中国驻德国大使史明德等人到机场迎接，不仅如此，机场还用高压水泵喷射形成水门，由飞机滑行通过。随后，大熊猫入住修葺一新的德国最古老的柏林动物园。

On June 24, 2017. Meng-Meng and Jiao-Qing the giant pandas arrived in Germany. Michael Müller, the Mayor of Berlin, and Shi Mingde, the Chinese Ambassador to Germany met the pandas at the airport. What was more, the airport also used high-pressure water pumps to form a water gate for the plane gliding through. Later, the pandas moved into Zoo Berlin, the oldest but renovated zoo in Germany.

“梦梦”和“娇庆”得到了悉心照料。2019年8月，“梦梦”顺利产下一对双胞胎“梦想”和“梦圆”。这是第一次有大熊猫在德国本土出生，弥补了德国人民的缺憾，实现了让“大熊猫幼仔出生在德国”的梦想，让德国人民感到无比欣喜和感动。

Meng-Meng and Jiao-Qing received carefully care. In August, 2019, Meng-Meng successfully gave birth to a twins, Meng-Xiang and Meng-Yuan. That was the first time that giant panda cubs were born in Germany, which patches up the shortcoming of the German people, who realized the dream of “giant panda cubs born in Germany”, and makes the German people feel very happy and moved.

大熊猫幼崽“梦圆”和“梦想”
giant panda cubs Meng-Xiang and Meng-Yuan
图片来源：柏林动物园推特截图 Berlin Zoo Twitter screenshot

2023年12月8日，德国柏林动物园为即将回中国的大熊猫双胞胎兄弟“梦想”和“梦圆”举办欢送活动，并为“梦想”“梦圆”纪念牌揭幕。活动当天，柏林中国文化中心还举办了以大熊猫为主题的中国文化体验活动，吸引了众多游客驻足观看并踊跃参与。2023年12月17日，“梦想”和“梦圆”回到中国成都。

On December 8, 2023, the Berlin Zoo in Germany held a farewell event and unveiled a commemorative plaque for the twin brothers, who would return to China. On the day of the event, the Chinese Cultural Center in Berlin held a Chinese cultural experience activity with the theme of giant pandas, attracting many tourists to watch and actively participate in it. On December 17, 2023, Meng-Xiang and Meng-Yuan returned to Chengdu, China.

“华豹”“金宝宝”踏雪芬兰

Hua-Bao & Jin-Baobao Snowshoed in Finland

2017年4月4日，中华人民共和国国家主席习近平对芬兰进行国事访问。期间，中国和芬兰签署了为期15年的大熊猫保护合作研究协议，以纪念这个欧洲国家独立100周年。2018年1月17日，大熊猫“华豹”和“金宝宝”启程前往多雪的芬兰。芬兰官员和中国驻芬兰大使在赫尔辛基万塔机场举行仪式欢迎大熊猫。

On April 4, 2017, Chinese President Xi Jinping paid a state visit to Finland. During his visit, China and Finland signed a 15-year agreement on giant panda conservation and joint research, marking the 100th anniversary of the independence of this European country. On January 17, 2018, Hua-Bao and Jin-Baobao the giant pandas left for snowy Finland. Finnish officials and the Chinese ambassador to Finland welcomed the pandas in a ceremony at the Helsinki’s airport.

为让大熊猫有一个良好的居住环境，芬兰艾赫泰里动物园耗资800万欧元建造了一个专门的熊猫屋。早在大熊猫抵芬之前，芬兰人就已经给中国熊猫起了芬兰名字：雄性熊猫“华豹”被命名为“派瑞”，雌性熊猫“金宝宝”被命名为“卢米”，分别是“雪”和“大雪”之意，这与赫尔辛基当日的天气相映成趣，因为就在大熊猫从成都启程的那一刻起，迟迟未降雪的芬兰首都赫尔辛基迎来了今冬第一场大雪。芬兰以特有的方式迎接大熊猫的到来。

In order to provide a good living environment for giant pandas, Ahtari Zoo had spent 8 million euros building a special Panda House. Early before the pandas arrived in Finland, the Finns had given the Chinese panda Finnish names: Pyry for the male panda Hua-Bao and Lumi for the female panda Jin-Baobao, which means Snow and Snowfall. This matched the weather in Helsinki that day because just at the moment when the giant panda started from Chengdu, Helsinki, the capital of Finland, received the first heavy snow of the winter. Finland welcomes the arrival of the pandas in a unique Finnish way.

"彩陶""湖春"存身印尼

Cai–Tao & Hu–Chun Stayed in Indonesia

2016年8月1日，在贵阳举行的中国-印尼副总理级人文交流机制第二次会议期间，在中华人民共和国国务院副总理刘延东和印尼人类发展与文化统筹部长普安的见证下，中国国家林业局与印尼环境与林业部签署了中国印尼共同推进大熊猫保护合作谅解备忘录。印尼成为与中国开展大熊猫研究合作的第16个国家。

On August 1, 2016, during the second meeting, held in Guiyang, of the China-Indonesia people-to-people exchange mechanism at vice premier's level, in the presence of Chinese Vice Premier Liu Yandong and Indonesian Coordinating Minister of Human Development and Culture Puan Maharani, The State Forestry Administration of China and the Kementerian LingKungan Hidup Dan Kehutanan (KLHK) of Indonesia signed the Memorandum of Understanding between China and Indonesia on jointly promoting cooperation on giant panda conservation. Indonesia became the 16th country to cooperate with China on giant panda research.

2017年9月28日，大熊猫“彩陶”和“湖春”前往印度尼西亚西爪哇省茂物县的野生动物园，这也是大熊猫首次旅居印尼。为迎接大熊猫到来，印尼特意在海拔1400米的山顶上修建了三层楼的熊猫馆，设立了大熊猫知识陈列室。

On September 28, 2017, Cai-Tao and Hu-Chun the giant pandas left for Taman Safari in Bogor County, West Java province, Indonesia. That was the first time giant pandas had lived in Indonesia. To welcome the arrival of the giant pandas, Indonesia has built a three-story panda house on the top of a mountain, located 1,400 meters above sea level, and set up a panda knowledge gallery.

印尼大熊猫馆
Panda House in Taman Safari, Indonesian

“丁丁”“如意”入住俄罗斯
Ding–Ding & Ru–Yi Settled in Russia

2019年2月28日，在中俄建交70周年之际，莫斯科动物园和中国野生动物保护协会签署了一项大熊猫保护研究合作协议，中方将租给俄方一对大熊猫以供双方合作开展大熊猫保护研究，期限为15年。同年4月26日，中华人民共和国国家主席习近平在北京会见俄罗斯总统普京时，宣布中国将向俄罗斯提供一对大熊猫开展合作研究。3天后，大熊猫“丁丁”“如意”抵达俄罗斯并于6月正式入住莫斯科动物园。

On February 28, 2019, on the occasion of the 70th anniversary of the establishment of diplomatic ties between China and Russia, Moscow Zoo and China Wildlife Conservation Association signed a cooperation agreement on giant panda conservation and research, in which China would loan Russia a pair of giant pandas for joint research, lasting 15 years. On April 26, when Chinese President Xi Jinping met with Russian President Vladimir Putin in Beijing, Xi Jinping once again confirmed that China will rent out giant pandas to Russia for cooperative research. Three days later, Ding-Ding and Ru-Yi the giant pandas arrived in Russia and moved into Moscow Zoo in June.

为迎接、安置好这对大熊猫，在大熊猫抵达莫斯科之前，俄罗斯已花费约十亿卢布建造了全新的莫斯科动物园熊猫馆。6月5日，中俄两国领导人共同出席了莫斯科动物园熊猫馆开馆仪式。目前，大熊猫在莫斯科动物园得到了良好的照料。

Before the giant pandas arrived in Moscow, to welcome and house the pandas, Russia had spent about 1 billion rubles in building the brand-new Panda House at the Moscow Zoo. On June 5, the heads of China and Russia jointly attended the opening ceremony of the Panda House at the Moscow Zoo. The pandas are now in good care at the Moscow Zoo.

2021年7月31日，俄罗斯莫斯科动物园游客为大熊猫“如意”庆祝生日
Tourists celebrate the birthday of the giant panda “Ruyi” at the Moscow Zoo
图片来源：https://www.sohu.com/a/480699508_267106
新华社发（叶甫盖尼·西尼岑　摄）

2023年4月，为深化中俄大熊猫保护研究合作，中国大熊猫保护研究中心专家应邀前往莫斯科动物园，为旅俄大熊猫“丁丁”“如意”繁殖工作提供技术指导。8月24日凌晨，“丁丁”顺利诞下雌性幼崽“喀秋莎”，这是在俄罗斯出生的首只大熊猫幼崽。大熊猫幼崽的诞生促进了中俄大熊猫科研合作，加深了两国人民的友谊，对俄罗斯和世界自然保护界来说都是具有里程碑意义的事件。

In April 2023, in order to deepen the cooperation between China and Russia in the protection and research of giant pandas, experts from the China Conservation and Research Center for the Giant Panda were invited to the Moscow Zoo to provide technical guidance for the breeding of Ding-Ding and Ru-Yi the giant pandas in Russia. Ding-Ding gave birth to a female cub Katyusha in the early morning of August 24. This is the first giant panda baby born in Russia. The birth of giant panda cub promotes the giant panda scientific research cooperation and the friendship between China and Russia, and is a milestone for both Russia and the world conservation community.

“星二”“毛二”奔赴丹麦
Xing-Er & Mao-Er Travelled to Denmark

2014年，丹麦女王玛格丽特二世来华进行国事访问期间，中丹两国就合作开展大熊猫保护研究达成共识。2017年5月3日，中国与丹麦正式签署大熊猫保护研究合作协议。2019年4月4日，大熊猫“星二”“毛二”飞往哥本哈根并顺利抵达凯斯楚普机场。2019年4月10日，哥本哈根动物园为大熊猫“星二”和“毛二”举行正式的欢迎仪式，女王玛格丽特二世出席并为熊猫馆剪彩。

During the Danish Queen’s state visit to China in 2014, China and Denmark reached consensus on giant pandas conservation and research. On May 3, 2017, China and Denmark officially signed the cooperative agreement on giant panda conservation and research. On April 4, 2019, Xing-Er and Mao-Er the giant pandas flew to Copenhagen and successfully arrived at the Copenhagen Kastrup Airport. On April 10, 2019, Copenhagen Zoo held an official welcome ceremony for the giant pandas Xing-Er and Mao-Er. Queen Margrethe Ⅱ attended and performed the ribbon-cutting for the panda pavilion.

4月11日上午，“熊猫与世界——中国大熊猫保护·文化·艺术成就展”在丹麦国家博物馆开幕，全面展现中国大熊猫保护、文化、艺术各方面取得的成绩。展览以沉浸式体验的方式，生动表现“星二”和“毛二”这两只大熊猫的成长过程，以及中丹两国大熊猫合作研究方面的最新进展。此外，还有十多位中外艺术创作者的优秀作品和大熊猫文创艺术品展出，共同为丹麦民众呈现了一幅美丽的中国大熊猫文化画卷。

On the morning of April 11, the achievement exhibition of“Panda and the World: Giant Panda Conversation Art Tour” opened at the National Museum of Denmark, fully showing the achievements of China's giant panda protection, culture and art. The exhibition vividly shows the growth process of the two giant pandas, Xing-Er and Mao-Er, as well as the research progress of the Chinese and Danish giant panda cooperation. At the same time, the excellent works of more than 10 Chinese and foreign art creators and the cultural and creative works of giant pandas have also opened a beautiful picture of Chinese giant panda culture for the Danish people.

丹麦大熊猫博物馆“太极宫”鸟瞰
Bird's eye view of Tai Chi Palace at the Danish Giant Panda Museum

为给大熊猫提供良好的居住环境，丹麦政府花费了1.5亿元人民币在丹麦动物园修建了大熊猫博物馆“太极宫”。该馆整体造型设计灵感来源于中国传统道家文化中的太极图案，充满了“阴阳”哲学。1958年，哥本哈根动物园曾短暂地养过一只熊猫“姬姬”，当时它在这里进行了三周的欧洲之旅。

In order to improve the living environment for giant pandas, the Danish government spent 150 million Yuan in building the giant panda museum Tai-Chi Palace in Copenhagen Zoo. The overall design is inspired by the Tai-Chi pattern in traditional Chinese Taoist culture, full of the philosophy of Yin and Yang. Copenhagen Zoo briefly had a panda back in 1958 when the panda Chi-Chi spent three weeks there as part of the tour to Europe.

“四海”“京京”亮相卡塔尔

Si-Hai & Jing-Jing Appeared in Qatar

为进一步促进大熊猫保护研究国际合作，2020年5月，中国和卡塔尔签订了大熊猫保护研究合作协议，以推动中卡两国濒危物种和生物多样性保护。2022年10月19日，大熊猫“四海”和“京京”抵达卡塔尔首都多哈并入住豪尔熊猫馆。

To further promote international cooperation on giant panda conservation and research, China and Qatar signed a cooperation agreement on giant panda conservation and research in May 2020 to promote the conservation of endangered species and biodiversity between China and Qatar. On October 19, 2022, Si-Hai and Jing-Jing the giant pandas arrived in Doha, capital of Qatar and merged into the Panda House in Al Khor Park.

出于对大熊猫的喜爱，卡塔尔将“京京”和“四海”以阿拉伯文分别命名为“苏海尔”和“索拉雅”。在阿拉伯传统文化中，“苏海尔”和“索拉雅”代表着吉祥、崇高和价值无限。11月17日，在2022卡塔尔足联世界杯即将开幕之际，“苏海尔”和“索拉雅”正式与公众见面。

Out of love for pandas, Qatar named the male panda Jing-Jing as Suhail and the female panda Si-Hai as Soraya in Arabic. In the traditional Arab culture, the names Suhail and Soraya represent auspiciousness, nobility and infinite value. On November 17, at the opening of the 2022 FIFA World Cup in Qatar, Suhail and Soraya officially met with the public.

卡塔尔民众参观大熊猫
Qatari people visit the giant pandas
中新社记者　富田　摄

卡塔尔气候干燥且天气炎热，这对熊猫的生活构成了巨大的挑战。为此，大熊猫馆通过精确控制湿度和温度，尽可能模拟大熊猫自然栖息地的四季变化，为生活在沙漠气候中的大熊猫提供一个凉爽的室内生活环境。

The dry climate and hot weather in Qatar pose a great challenge to the life of the pandas. To this end, the giant panda house accurately controls the humidity and temperature to simulate the seasonal changes of the natural habitat of giant pandas as much as possible, providing cool indoor living conditions for giant pandas in the desert climate.

“金喜”“茱萸”前往西班牙

Jin–Xi & Zhu–Yu Arrived Spain

2024年4月29日，中国野生动物保护协会宣布，根据中国大熊猫国际保护合作工作总体部署，来自成都大熊猫繁育研究基地的大熊猫“金喜”“茱萸”即将前往西班牙马德里动物园，接替大熊猫“花嘴巴”一家，开启为期10年的旅居生活。这也是自2019年下半年以来首批飞往欧洲旅居的大熊猫。

On April 29,2024, the CWCA announced that according to the overall deployment of the international cooperation on giant panda protection of China, giant pandas Jin-Xi and Zhu-Yu from Chengdu Research Base of Giant Panda Breeding would go to the Madrid Zoo, Spain, to replace the giant panda Hua-Zuiba family to start a 10-year sojourn life. It is also the first batch of pandas to fly to Europe since the second half of 2019.

2024年4月29日上午，大熊猫“金喜”“茱萸”从成都双流国际机场飞往西班牙。在经过一个月的隔离检疫和适应期后，当地时间2024年5月30日，大熊猫正式与西班牙公众见面。西班牙王太后索菲亚出席并主持了亮相仪式。

On the morning of April 29,2024, giant pandas Jin-Xi and Zhu-Yu flew from Chengdu Shuangliu International Airport in Chengdu to Spain. After a one-month quarantine and adaptation period, the giant pandas officially met with the Spanish public on May 30,2024. Queen Dowager Sophia of Spain attended and presided over the ceremony.

作为“花嘴巴”一家的接替者，大熊猫“金喜”“茱萸”将在新的环境中继续传递爱与和平，促进两国的友谊和文化交流。中国驻西班牙大使姚敬指出，这些非常特别的动物已经成为中国和西班牙友谊的重要象征。

As the replacement of the Hua-Zuiba family, Jin-Xi and Zhu-Yu will continue to convey the love and peace in the new environment, and promote the friendship and cultural exchange between the China and Spain. Yao Jing, the Chinese ambassador to Spain, pointed out that these very special animals have become an important symbol of the friendship between China and Spain.

“云川”“鑫宝”接棒美国

Yun-Chuan & Xin-Bao Relieved in USA

2024年4月27日，国家林业和草原局宣布，中国野生动物保护协会与美国圣迭戈动物园签署了为期10年的大熊猫保护合作协议。将选择一对大熊猫前往圣迭戈动物园，开启新一轮大熊猫国际保护合作。

On April 27,2024, the National Forestry and Grassland Administration of China announced that the CWCA and San Diego Zoo signed a 10-year cooperation agreement on the protection of giant pandas, sending a pair of giant pandas to San Diego Zoo to launch a new round of international cooperation on the protection of giant pandas.

“云川”“鑫宝”欢送仪式
Farewell Ceremony for Yun-Chuan and Xin-Bao

2024年6月26日晚，大熊猫“云川”和“鑫宝”从中国保护大熊猫研究中心雅安碧峰峡基地启程前往美国并于次日抵达圣迭戈动物园。

On the evening of June 26,2024, giant pandas Yun-Chuan and Xin-Bao departed from the Bifengxia Base of the Giant Panda Conservation and Research Center in Ya'an for the United States and arrived at the San Diego Zoo the next day.

圣迭戈动物园是美国首家与中国开展大熊猫合作研究的机构，也是全球规模最大、享有盛誉的世界知名动物园之一。自1994年开展合作以来，中国保护大熊猫研究中心与圣迭戈动物园双方在大熊猫生态学、行为学、遗传学、营养学等领域研究，以及野外监测、人工繁育、疾病防治、伴生物种研究等方面取得了丰硕成果，共同攻克了系列技术难题，为全球濒危野生动物保护做出了贡献。

San Diego Zoo is the first institution in the United States to conduct giant panda research with China. It is also one of the largest and world-renowned zoos in the world. Since 1994, the China Conservation and Research Center for the Giant Panda and San Diego zoo have achieved fruitful results in the giant panda ecology, ethology , genetics, nutriology, as well as the field monitoring, artificial breeding, disease prevention, associated species research, jointly overcame a series of technical problems, contributed to the global endangered wildlife protection.

附录1 短期出国巡展大熊猫统计表

Appendix 1 Statistics of Giant Pandas for Short-term Overseas Tour Exhibitions

序号 No.	谱系号 Stud No.	中文名	Name	出国巡展日期 Date of Going Abroad	展出地 Residence
1	59	珊珊	Hsan-Hsan	1980.03—1980.06	日本福冈Japan
2	77	宝玲	Bao-Ling	1980.03—1980.06	日本福冈Japan
3	155	伟伟	Wey-Wey	1981.02—1981.03 1986.02—1986.06 1991.06—1992.03	日本Japan 日本Japan 印尼Indonesia
4	163	蓉蓉	Rong-Rong	1981.03—1981.09	日本神户Japan
5	195	寨寨	Zhai-Zhai	1981.03—1981.09	日本神户Japan
6	242	迎新	Ying-Xin	1984.07—1984.10 1984.10—1985.01	美国洛杉矶动物园USA 美国旧金山动物园USA
7	245	永永	Yong-Yong	1984.07—1984.10 1984.10—1985.01 1987.04—1987.11 1987.11—1988.04	美国洛杉矶动物园USA 美国旧金山动物园USA 美国纽约布朗克斯动物园USA 美国布什公园USA
8	222	青青	Qing-Qing	1985.07—1985.09	加拿大多伦多动物园Canada
9	251	全全	Quan-Quan	1985.07—1985.10	加拿大多伦多动物园Canada
10	202	川川	Chuan-Chuan	1986.04—1986.08 1987.05—1987.09 1992.12—1993.06	瑞典Sweden 荷兰卑尔根野生动物园Holland 泰国曼谷野生动物园Thailand
11	209	平平	Ping-Ping	1986.05—1986.10	爱尔兰都柏林动物园Ireland
12	214	明明	Ming-Ming	1986.05—1986.10 1991.10—1994.10	爱尔兰都柏林动物园Ireland 英国伦敦动物园UK

续 表

序号 No.	谱系号 Stud No.	中文名	Name	出国巡展日期 Date of Going Abroad	展出地 Residence
13	294	陵陵	Ling-Ling	1987.04—1987.11 1987.11—1988.10 2001.01—2001.04	美国纽约布朗克斯动物园USA 美国布什公园USA 墨西哥动物园Mexico
14	312	苏苏	Su-Su	1987.05—1987.09	荷兰卑尔根野生动物园Holland
15	296	希希	Xi-Xi	1988.02—1988.09	加拿大卡尔加里动物园Canada
16	191	弯弯	Wan-Wan	1987.06—1987.09	比利时安特卫普动物园 Belgium
17	330	习习	Xi-Xi	1987.06—1987.09	比利时安特卫普动物园 Belgium
18	264	巴斯	Ba-Si	1987.07—1988.02	美国圣迭戈动物园USA
19	282	元元	Yuen-Yuen	1987.07—1988.02	美国圣迭戈动物园USA
20	219	威伦	Wei-Lun	1988.02—1988.09	加拿大卡尔加里动物园Canada
21	384	娇娇	Jiao-Jiao	1988.02—1988.03	新加坡动物园Singapore
22	271	南南	Nan-Nan	1988.03—1988.11	美国托莱多动物园USA
23	305	乐乐	Lo-Lo	1988.03—1988.11	美国托莱多动物园USA
24	278	庆庆	Qing-Qing	1988.03—1988.07 1988.07—1988.10 1988.10—1989.01	日本池田动物园Japan 日本涵馆市动物园Japan 日本和歌山野生动物园Japan
25	297	珍珍	Chen-Chen	1988.03—1988.07 1988.07—1988.10 1988.10—1989.01 1989.03—1989.09	日本池田动物园Japan 日本涵馆市动物园Japan 日本和歌山野生动物园Japan 加拿大温尼伯动物园Canada
26	283	菲菲	Fei-Fei	1988.03—1988.07 1988.07—1988.10 1988.10—1989.01	澳大利亚墨尔本动物园 Australia 澳大利亚悉尼动物园Australia 新西兰奥克兰动物园 New Zealand

续 表

序号 No.	谱系号 Stud No.	中文名	Name	出国巡展日期 Date of Going Abroad	展出地 Residence
27	290	小小	Xiao-Xiao	1988.03—1988.07 1988.10—1989.01	澳大利亚墨尔本动物园 Australia 新西兰奥克兰动物园 New Zealand
28	300	刚刚	Gang-Gang	1989.03—1989.09	加拿大温尼伯动物园Canada
29	314	冰冰	Bing-Bing	1989.03—1989.09	加拿大温尼伯动物园Canada
30	327	安安	An-An	1990.09—1991.01	新加坡动物园Singapore
31	329	新兴	Xin-Xing	1990.09—1991.01	新加坡动物园Singapore
32	356	星星	Xing-Xing	1991.02—1992.04	美国哥伦布动物园USA
33	364	亚庆	Yia-Qing	1992.12—1993.06	泰国曼谷野生动物园Thailand
34	247	文文	Wen-Wen	2001.06—2001.08	俄罗斯莫斯科动物园Russia
35	450? 488?	奔奔? 园园?	Ben-Ben? Yuan-Yuan?	2001.06—2001.08	俄罗斯莫斯科动物园Russia

注：本表根据《大熊猫谱系（2017）》整理，可能不完整，个别可能存在疑问，比如，谱系记载2001年“园园”曾短暂旅居莫斯科动物园，但相关资料表明，2001年“文文”和“奔奔”赴莫斯科参加文化周活动，助力中国申奥，而谱系中却没有“奔奔”赴莫斯科的信息。还有资料表明，1981年重庆动物园大熊猫“松松”和“南南”应邀在加拿大各地巡游半年多，同年北京动物园大熊猫“迎新”和“媛媛”曾出访安哥拉，这些在谱系中亦无任何记录。

附录2 出国旅居大熊猫统计表

Appendix 2 Statistics of Giant Pandas Living Abroad

序号 No.	谱系号 Stud No.	中文名	Name	性别 Sex	出国日期 Date of Going Abroad	生卒日期 Date of Birth & Death	旅居地Residence
1	1	苏琳	Su-Lin	雄M	1936.12.02	约1936.09.00-1938.04.01	美国芝加哥动物园 Chicago Zoological Park, USA
2	2	珍妮	Jennie	雌F	1937.07.00	约1936.00.00-1937.07.00	死于赴英国途中 On the way to England
3	3	妹妹（戴安娜）	Mei-Mei（Diana）	雄M	1938.01.28	约1936.00.00-1942.08.03	美国纽约布朗克斯动物园 Bronx Zoo, USA
4	4	潘多拉	Pandora	雌F	1938.05.18	约1937.00.00-1941.05.13	美国纽约布朗克斯动物园 Bronx Zoo, USA
5	5	开心果	Happy	雄M	1938.12.24	约1935.09.00-1946.03.10	英国伦敦动物园 London Zoo, UK
6	6	唐	Tang	雄M	1938.12.24	约1936.00.00-1940.04.23	英国伦敦动物园 London Zoo, UK
7	7	宋	Sung	雄M	1938.12.24	约1936.00.00-1939.12.18	英国伦敦动物园 London Zoo, UK
8	8	奶奶	Grandma	雌F	1938.12.24	约1936.00.00-1939.01.09	英国伦敦动物园 London Zoo, UK
9	9	明	Ming	雌F	1938.12.24	约1937.00.00-1944.12.26	英国伦敦动物园 London Zoo, UK
10	10	美兰	Mei-Lan	雄M	1939.11.16（Arrival）	约1938.00.00-1953.09.05	美国芝加哥动物园 Chicago Zoological Park, USA
11	12	宝贝	Pao-Pei	雌F	1939.02.00	约1938.09.00-1940.10.04	美国圣路易斯动物园 St. Louis Zoo, USA
12	11	潘	Pan	雄M	1939.05.01	约1938.00.00-1940.05.05	美国纽约布朗克斯动物园 Bronx Zoo, USA

续表

序号 No.	谱系号 Stud No.	中文名	Name	性别 Sex	出国日期 Date of Going Abroad	生卒日期 Date of Birth & Death	旅居地Residence
13	13	潘弟	Pan-Dee	雌F	1941.11.09	约1940.00.00-1945.10.04	美国纽约布朗克斯动物园 Bronx Zoo, USA
14	14	潘达	Pan-Dah	雄M	1941.11.09	约1940.00.00-1951.10.31	美国纽约布朗克斯动物园 Bronx Zoo, USA
15	15	联合	Lien-Ho（Unity）	雌F	1946.05.01	约1945.00.00-1950.02.22	英国伦敦动物园 London Zoo, UK
16	17	平平	Ping-Ping	雄M	1957.05.18	约1953.00.00-1961.05.29	苏联莫斯科动物园 Moscow Zoo, U.S.S.R
17	18	碛碛（姬姬）	Qi-Qi（Chi-Chi）	雌F	1957.05.18/1958.09.26	约1954.09.00-1972.07.22	英国伦敦动物园 London Zoo, UK
18	24	安安	An-An	雄M	1959.08.18	约1956.08.00-1972.10.18	苏联莫斯科动物园 Moscow Zoo, U.S.S.R
19	70	一号	Number 1	雌F	1965.06.03	????.??.??-1971.10.20	朝鲜中央动物园 Central Zoo, DPRK
20	71	二号	Number 2	雄M	1965.06.03	????.??.??-????.??.??	朝鲜中央动物园 Central Zoo, DPRK
21	101	凌凌	Lin-Lin	雄M	1971.10.20	1969.09.00-1972.10.18	朝鲜中央动物园 Central Zoo, DPRK
22	107	三星	San-Xing	雌F	1971.10.20	1968.09.00-1988.00.00	朝鲜中央动物园 Central Zoo, DPRK
23	133	丹丹	Dan-Dan	雄M	1979.03.20	1970.09.00-1988.00.00	朝鲜中央动物园 Central Zoo, DPRK
24	112	玲玲	Ling-Ling	雌F	1972.04.16	1970.09.00-1992.12.30	美国国家动物园 National Zoological Park, USA
25	121	兴兴	Hsing-Hsing	雄M	1972.04.16	1971.08.00-1999.11.28	美国国家动物园 National Zoological Park, USA

续表

序号 No.	谱系号 Stud No.	中文名	Name	性别 Sex	出国日期 Date of Going Abroad	生卒日期 Date of Birth & Death	旅居地Residence
26	111	兰兰	Lan-Lan	雌F	1972.10.28	1969.00.00-1979.09.04	日本上野动物园 Ueno Zoo, Japan
27	122	康康	Kang-Kang	雄M	1972.10.28	1971.08.00-1980.06.30	日本上野动物园 Ueno Zoo, Japan
28	138	黎黎	Li-Li	雄M	1973.12.08	1972.09.00-1974.04.20	法国文森动物园 Vincennes Zoological Park, France
29	140	燕燕	Yan-Yan	雄M	1973.12.08	1972.09.00-2000.01.20	法国文森动物园 Vincennes Zoological Park, France
30	127	晶晶	Ching-Ching	雌F	1974.09.14	1972.09.00-1985.06.20	英国伦敦动物园 London Zoo, UK
31	141	佳佳	Chia-Chia	雄M	1974.09.14	1972.09.00-1991.10.13	英国伦敦动物园 London Zoo, UK
32	165	迎迎	Ying-Ying	雌F	1975.09.10	1974.09.00-1989.01.29	墨西哥查普特佩克动物园 Chapultepec Zoo, Mexico
33	167	贝贝	Pe-Pe	雄M	1975.09.10	1974.09.00-1988.07.20	墨西哥查普特佩克动物园 Chapultepec Zoo, Mexico
34	169	绍绍	Shao-Shao	雌F	1978.12.28	1975.09.18-1983.10.23	西班牙马德里动物园 Zoo Aquarium de Madrid, Spain
35	187	强强	Chang-Chang	雄M	1978.12.28	1971.00.00-1995.10.13	西班牙马德里动物园 Zoo Aquarium de Madrid, Spain
36	162	欢欢	Huan-Huan	雌F	1980.01.29	1973.00.00-1997.09.21	日本上野动物园 Ueno Zoo, Japan
37	208	飞飞	Fei-Fei	雄M	1982.10.09	1967.09.00-1994.12.14	日本上野动物园 Ueno Zoo, Japan
38	210	陵陵	Ling-Ling	雄M	1992.11.05	1985.09.05-2008.04.30	日本上野动物园 Ueno Zoo, Japan

续表

序号 No.	谱系号 Stud No.	中文名	Name	性别 Sex	出国日期 Date of Going Abroad	生卒日期 Date of Birth & Death	旅居地Residence
39	183	宝宝	Bao-Bao	雄M	1980.11.05	1978.09.00-2012.08.22	德国柏林动物园 Zoo Berlin, West Germany
40	294	天天	Tian-Tian	雌F	1980.11.05	1978.00.00-1984.02.08	德国柏林动物园 Zoo Berlin, West Germany
41	378	嫣嫣	Yan-Yan	雌F	1995.00.00	1985.08.00-2007.03.26	德国柏林动物园 Zoo Berlin, West Germany
42	389	蓉浜	Rong-Bin	雌F	1994.09.06	1992.09.04-1997.07.17	日本和歌山动物园 Adventure World in Shirahama, Japan
43	390	永明	Yong-Ming	雄M	1994.09.06	1992.09.14-	日本和歌山动物园 Adventure World in Shirahama, Japan
44	408	梅梅	Mei-Mei	雌F	2000.07.07	1994.08.31-2008.10.15	日本和歌山动物园 Adventure World in Shirahama, Japan
45	385	明明（川星）	Ming-Ming（Chuan-Xing）	雌F	1994.09.20	1992.07.18-2016.09.09	韩国三星爱宝乐园 Samsung Everland, ROK
46	387	莉莉	Li-Li	雌F	1994.09.20	1992.09.03-	韩国三星爱宝乐园 Samsung Everland, ROK
47	371	白云	Bai-Yun	雌F	1996.09.10	1991.09.07-	美国圣迭戈动物园 San Diego Zoo, USA
48	381	石石	Shi-Shi	雄M	1996.09.10	1982.00.00-2008.07.05	美国圣迭戈动物园 San Diego Zoo, USA
49	452	高高	Gao-Gao	雄M	2003.01.14	1992.04.06-	美国圣迭戈动物园 San Diego Zoo, USA
50	461	伦伦（华华）	Lun-Lun	雌F	1999.11.05	1997.08.25-	美国亚特兰大动物园 Zoo Atlanta, USA
51	415	洋洋（九九）	Yang-Yang	雄M	1999.11.05	1997.09.09-	美国亚特兰大动物园 Zoo Atlanta, USA

续表

序号 No.	谱系号 Stud No.	中文名	Name	性别 Sex	出国日期 Date of Going Abroad	生卒日期 Date of Birth & Death	旅居地Residence
52	434	爽爽（旦旦）	Shuang-Shuang	雌F	2000.07.16	1995.09.16-	日本神户王子动物园 Kobe Oji Zoo, Japan
53	437	锦竹（兴兴1）	Jin-Zhu（Xing-Xing 1）	雌F	2000.07.16	1996.08.12-2014.10.14	日本神户王子动物园 Kobe Oji Zoo, Japan
54	433	龙龙（兴兴2）	Long-Long（Xing-Xing 2）	雄M	2002.12.09	1995.09.14-2010.09.09	日本神户王子动物园 Kobe Oji Zoo, Japan
55	458	添添	Tian-Tian	雄M	2000.12.06	1997.08.27-2023.11.09	美国国家动物园 National Zoological Park，USA
56	473	美香	Mei-Xiang	雌F	2000.12.06	1998.07.22-2023.11.09	美国国家动物园 National Zoological Park，USA
57	514	阳阳	Yang-Yang	雌F	2003.03.03	2000.08.10-	奥地利美泉宫动物园 Tiergarten Schönbrunn, Austria
58	526	龙徽	Long-Hui	雄M	2003.03.03	2000.09.26-2016.12.09	奥地利美泉宫动物园 Tiergarten Schönbrunn, Austria
59	488	园园	Yuan-Yuan	雄M	2019.04.15	1999.08.23-	奥地利美泉宫动物园 Tiergarten Schönbrunn, Austria
60	466	乐乐	Le-Le	雄M	2003.04.07	1998.07.18-2023.02.01	美国孟菲斯动物园 Memphis Zoo, USA
61	507	丫丫	Ya-Ya	雌F	2003.04.07	2000.08.03-	美国孟菲斯动物园 Memphis Zoo, USA
62	510	创创	Chuang-Chuang	雄M	2003.10.12	2000.08.06-2019.09.16	泰国清迈动物园 Chiang Mai Zoo, Thailand
63	539	林惠	Lin-Hui	雌F	2003.10.12	2001.09.28-2023.04.19	泰国清迈动物园 Chiang Mai Zoo, Thailand
64	519	冰星	Bing-Xing	雄M	2007.09.08	2000.09.01-	西班牙马德里动物园 Zoo Aquarium de Madrid, Spain

续表

序号 No.	谱系号 Stud No.	中文名	Name	性别 Sex	出国日期 Date of Going Abroad	生卒日期 Date of Birth & Death	旅居地Residence
65	576	花嘴巴	Hua-Zuiba	雌F	2007.09.08	2003.09.16-	西班牙马德里动物园 Zoo Aquarium de Madrid, Spain
66	620	网网	Wang-Wang	雄M	2009.11.27	2005.08.31	澳大利亚阿德莱德动物园 Adelaide Zoo，Australia
67	638	福妮	Fu-Ni	雌F	2009.11.27	2006.08.23-	澳大利亚阿德莱德动物园Adelaide Zoo，Australia
68	600	仙女（真真）	Xian-Nü	雌F	2011.02.21	2005.07.03-	日本上野动物园 Ueno Zoo, Japan
69	612	比力（力力）	Bi-Li	雄M	2011.02.21	2005.08.16-	日本上野动物园 Ueno Zoo, Japan
70	564	阳光	Yang-Guang	雄M	2011.12.04	2003.08.14-	英国爱丁堡动物园 Edinburgh Zoo, UK
71	569	甜甜	Tian-Tian	雌F	2011.12.04	2003.08.24-	英国爱丁堡动物园 Edinburgh Zoo, UK
72	723	欢欢	Huan-Huan	雌F	2012.01.15	2008.08.10-	法国博瓦尔动物园 ZooParc de Beauval, France
73	736	圆仔	Yuan-Zi	雄M	2012.01.15	2008.09.06-	法国博瓦尔动物园 ZooParc de Beauval, France
74	690	武杰（凯凯）	Wu-Jie	雄M	2012.09.05	2007.09.14-	新加坡河川生态园 River Safari, Singapore
75	734	沪宝（嘉嘉）	Hu-Bao	雌F	2012.09.05	2008.09.03-	新加坡河川生态园 River Safari, Singapore
76	676	二顺	Er-Shun	雌F	2013.03.25	2007.08.10-	加拿大多伦多动物园/ Toronto Zoo, Canada 卡尔加里动物园 Calgary Zoo, Canada

续表

序号 No.	谱系号 Stud No.	中文名	Name	性别 Sex	出国日期 Date of Going Abroad	生卒日期 Date of Birth & Death	旅居地Residence
77	732	大毛	Da-Mao	雄M	2013.03.25	2008.09.01-	加拿大多伦多动物园/ Toronto Zoo, Canada 卡尔加里动物园 Calgary Zoo, Canada
78	741	好好	Hao-Hao	雌F	2014.02.22	2009.07.07-	比利时天堂动物园 PairiDaiza, Belgium
79	745	星徽	Xing-Hui	雄M	2014.02.22	2009.07.22-	比利时天堂动物园 PairiDaiza, Belgium
80	639	福娃（兴兴）	Fu-Wa	雄M	2014.05.21	2006.08.23-	马来西亚国家动物园 National Zoo & Aquarium, Malaysia
81	641	凤仪（靓靓）	Feng-Yi	雌F	2014.05.21	2006.08.23-	马来西亚国家动物园 National Zoo & Aquarium, Malaysia
82	841	园欣（乐宝）	Yuan-Xin	雄M	2016.03.03	2012.07.28-	韩国三星爱宝乐园 Samsung Everland，ROK
83	869	华妮（爱宝）	Hua-Ni	雌F	2016.03.03	2013.07.13-	韩国三星爱宝乐园 Samsung Everland，ROK
84	879	星雅	Xing-Ya	雄M	2017.04.12	2013.08.05-	荷兰欧维汉动物园 Ouwehands Dierenpark, Netherlands
85	884	武雯	Wu-Wen	雌F	2017.04.12	2013.08.11-	荷兰欧维汉动物园 Ouwehands Dierenpark, Netherlands
86	769	娇庆	Jiao-Qing	雄M	2017.06.23	2010.07.15-	德国柏林动物园 Zoo Berlin, Germany
87	868	梦梦	Meng-Meng	雌F	2017.06.24	2013.07.10-	德国柏林动物园 Zoo Berlin, Germany
88	778	彩陶	Cai-Tao	雄M	2017.09.28	2010.08.04-	印尼野生动物园 Taman Safari Bogor, Indonesia
89	794	湖春	Hu-Chun	雌F	2017.09.28	2010.09.08	印尼野生动物园 Taman Safari Bogor, Indonesia

续表

序号 No.	谱系号 Stud No.	中文名	Name	性别 Sex	出国日期 Date of Going Abroad	生卒日期 Date of Birth & Death	旅居地Residence
90	941	金宝宝	Jin-Baobao（Lumi）	雌F	2018.01.17	2014.09.20-	芬兰阿塔利动物园 Ähtäri Zoo, Finland
91	949	华豹	Hua-Bao（Pyry）	雄M	2018.01.17	2015.06.25-	芬兰阿塔利动物园 Ähtäri Zoo, Finland
92	900	星二	Xing-Er	雄M	2019.04.07	2013.08.23-	丹麦哥本哈根动物园 Copenhagen Zoo, Denmark
93	919	毛二	Mao-Er	雌F	2019.04.07	2014.07.26-	丹麦哥本哈根动物园 Copenhagen Zoo, Denmark
94	1021	如意	Ru-Yi	雄M	2019.04.29	2016.07.31-	俄罗斯莫斯科动物园 Moscow Zoo, Russia
95	1087	丁丁	Ding-Ding	雌F	2019.04.29	2017.07.30-	俄罗斯莫斯科动物园 Moscow Zoo, Russia
96	1166	京京（苏海尔）	Jing-Jing (Suhail)	雄M	2022.10.19	2018.09.19-	卡塔尔豪尔公园 Al Khor Park, Qatar
97	1191	四海（索拉雅）	Si-Hai (Thuraya)	雌F	2022.10.19	2019.07.26-	卡塔尔豪尔公园 Al Khor Park, Qatar
98	1264	金喜	Jin-Xi	雄M	2024.04.29	2020.09.01-	西班牙马德里动物园 Zoo Aquarium de Madrid, Spain
99	1252	茱萸	Zhu-Yu	雌F	2024.04.29	2020.10.25-	西班牙马德里动物园 Zoo Aquarium de Madrid, Spain
100	1193	云川	Yun-Chuan	雄M	2024.06.26	2019.07.28-	美国圣迭戈动物园 San Diego Zoo,USA
101	1286	鑫宝	Xin-Bao	雌F	2024.06.26	2020.07.23-	美国圣迭戈动物园 San Diego Zoo,USA

参考文献

[1] [美]George B. Schaller. *The Last Panda*[M]. The University of Chicago Press, Ltd., 1993.

[2] [美]Helen M. Fox. *Abbe David's Diary*[M]. Harvard University Press,2014.

[3] [美]Theodore Roosevelt, Kermit Roosevelt. *Trailing The Giant Panda*[M]. Blue Ribbon Books,1929.

[4] [美]Ruth Harkness. *The Lady and the Panda: An Adventure*[M]. Carrick & Evans, 1938.

[5] [美]Vicki Croke. *The Lady and the Panda: The True Adventures of the First American Explorer to Bring Back China's Most Exotic Animal*[M].Random House,2005.

[6] [美]乔治·夏勒．最后的熊猫[M]．张定绮，译，胡锦矗，校．上海：上海译文出版社，2015.

[7] [美]西奥多·罗斯福，克米特·罗斯福．跟踪大熊猫的足迹[M]．昆明：云南民族出版社，2014.

[8] [美]克鲁克．淑女与熊猫[M]．苗华建，译，北京：新星出版社，2007.

[9] [英]亨利・尼科尔斯．来自中国的礼物：大熊猫与人类相遇的一百年[M]．黄建强，译，北京：生活・读书・新知三联书店，2018.

[10] 考拉看看．大熊猫之路：一部绚烂的大熊猫文明史[M]．北京：现代出版社，2023.

[11] 高富华．大熊猫，国宝的百年传奇[M]．北京：五洲传播出版社，2020.

[12] 高富华．大熊猫史话[M]．成都：四川民族出版社，2019.

[13] 孙前．大熊猫文化笔记[M]．北京：中国发展出版社，2019.

[14] 赵良冶．熊猫中国：中国大熊猫纪实[M]．南京：江苏凤凰文艺出版社，2019.

[15] 少君，平萍．“美女”与熊猫：一个真实的故事[M]．成都：成都时代出版社，2006.

[16] 黄贵璋，桂明义，胡大可．宝兴大熊猫[M]．四川省宝兴县政协文史委内部资料，1994.

[17] 杨玉君．熊猫老家・国宝档案[M]．四川蜂桶寨国家级自然保护区管护中心内部资料，2021.

[18] 姜鸿．科学、商业与政治：走向世界的中国大熊猫（1869—1948）[J]．近代史研究，2021（01）.

[19] 张恩铭．简述20世纪50—70年代中国的动物外交活动[J]．惠州学院学报，2018（04）.

[20] 黄金生．大熊猫出国史话[J]．文史精华，2017（20）.

[21] 薇薇．中国“熊猫外交”往事[J]．党的生活（河南），2016（08）.

[22] 刘晓晨．兄弟之盟：新中国同社会主义国家间的动物外交[J]．史林，2015（02）.

[23] 李果，余佩妮．中马“熊猫外交”[J]．华商，2014（10）.

[24] 甘子玲．熊猫外交：1957—1982没有回程票的熊猫大使[J]．看世界，2014（11）．

[25] 邵铭煌．抗战时期鲜为人知的“熊猫外交”[J]．抗战史料研究，2012（01）．

[26] 杨程屹．熊猫外交记[J]．看历史，2011（04）．

[27] 刘庶光．忆为尼克松选送熊猫的一段往事[J]．晚霞，2011（19）．

[28] 林禾，余里．中国“熊猫外交”揭秘[J]．传承，2010（02）．

[29] 静水．中国大熊猫的海外旧事[J]．档案与史学，2010（11）．

[30] 唐润明．宋美龄与熊猫外交[J]．兰台世界，2009（03）．

[31] 孙佳华，夏俊．“熊猫外交”的秘密往事[J]．政府法制，2009（21）．

[32] 鲁银．旅居海外的“熊猫使者”[J]．地图，2006（03）．

[33] 钟肇敏．大熊猫首赴爱尔兰[J]．纵横，2000（09）．

后记

数百万年来，大熊猫始终迈着雍容的步伐，追星逐月。它们带着独有的灵气，走过数百万年的时光，最终与人类相遇。此后，它们的身份经历了多次转变，从自由自在的竹林隐士到被视为奇珍异兽，从作为至尊国礼到成为科研助手。

作为中国的国宝，大熊猫在推动中外文化交融和人文交流中发挥了重要作用，被视为友谊的使者和桥梁。本书聚焦于大熊猫文化的国际传播，旨在为对大熊猫及其文化感兴趣的中外青少年及学者提供参考。

本书基于对大熊猫首次科学发现至今参与国际交流情况的系统梳理和全面考证，对相关大熊猫进行了简要介绍。内容史料性较强，具有较大的参考价值。然而，由于作者水平所限，加之部分史料时间久远，难以完全收集齐全，书中难免存在不足之处。欢迎读者提出宝贵意见，以便后续修订和完善。

Postscript

For millions of years, the giant pandas have always been taking a graceful pace, chasing stars and moon. Through millions of years, they meet with humans with a unique nimbus. Since then, they have experienced the transformation from free bamboo-forest hermits to rare species, from the supreme state gifts to scientific research assistants.

As national treasure of China, giant panda have played an important role in promoting the cultural integration and culturala Exchanges between China and foreign countries, and is regarded as a messenger and bridge of friendship. This book focuses on the international communication of giant panda culture, aiming to provide reference for Chinese and foreign teenagers and scholars who love the giant panda and giant panda culture to understand the giant panda and giant panda culture.

On the basis of systematic review and comprehensive research, this book briefs the giant pandas involved in international communication since the first scientific discovery, and fulfills rich historical data and great reference value. Opinions are welcome for improvement.

致 谢

Acknowledgments

本书得到了大熊猫文化研究专家、雅安市社会科学界联合会副主席杨铧先生的不吝指教与大力支持，在此表示衷心的感谢。

Special thanks to Mr. Yang Hua, a researcher of giant panda culture, and the vice-chairman of Ya’an Social Sciences Association.